**Springer Theses**

Recognizing Outstanding Ph.D. Research

## Aims and Scope

The series "Springer Theses" brings together a selection of the very best Ph.D. theses from around the world and across the physical sciences. Nominated and endorsed by two recognized specialists, each published volume has been selected for its scientific excellence and the high impact of its contents for the pertinent field of research. For greater accessibility to non-specialists, the published versions include an extended introduction, as well as a foreword by the student's supervisor explaining the special relevance of the work for the field. As a whole, the series will provide a valuable resource both for newcomers to the research fields described, and for other scientists seeking detailed background information on special questions. Finally, it provides an accredited documentation of the valuable contributions made by today's younger generation of scientists.

## Theses are accepted into the series by invited nomination only and must fulfill all of the following criteria

- They must be written in good English.
- The topic should fall within the confines of Chemistry, Physics, Earth Sciences, Engineering and related interdisciplinary fields such as Materials, Nanoscience, Chemical Engineering, Complex Systems and Biophysics.
- The work reported in the thesis must represent a significant scientific advance.
- If the thesis includes previously published material, permission to reproduce this must be gained from the respective copyright holder.
- They must have been examined and passed during the 12 months prior to nomination.
- Each thesis should include a foreword by the supervisor outlining the significance of its content.
- The theses should have a clearly defined structure including an introduction accessible to scientists not expert in that particular field.

Indexed by zbMATH.

More information about this series at http://www.springer.com/series/8790

Kuan Hu

# Development of In-Tether Carbon Chiral Center-Induced Helical Peptide

## Methodology and Applications

Doctoral Thesis accepted by
Peking University, Beijing, China

*Author*
Dr. Kuan Hu
School of Chemical Biology and
Biotechnology, Peking University
Shenzhen Graduate School
Shenzhen, China

*Supervisor*
Prof. Zigang Li
School of Chemical Biology and
Biotechnology, Peking University
Shenzhen Graduate School
Shenzhen, China

ISSN 2190-5053 ISSN 2190-5061 (electronic)
Springer Theses
ISBN 978-981-33-6612-1 ISBN 978-981-33-6613-8 (eBook)
https://doi.org/10.1007/978-981-33-6613-8



This Springer imprint is published by the registered company Springer Nature Singapore Pte Ltd.
The registered company address is: 152 Beach Road, #21-01/04 Gateway East, Singapore 189721, Singapore

# Supervisor's Foreword

At the beginning of the twenty-first century, chemical biologists have already begun to use chemical modification methods to introduce a cross-linker into a peptide to obtain a stapled peptide with a regular helical conformation. Therefore, the use of chemically modified stabilized peptides has become a research focal point, and related applications involve a series of important biological research fields such as protein-protein interactions. However, for many years, the mechanism of how the helical structure affects the properties of a peptide has been difficult to be clearly explained. In his thesis, Dr. Kuan Hu keenly discovered that when a chiral center was introduced into the cross-linker of a stapled peptide, the configuration of the chiral center largely determines the helical content of a peptide. In other words, by adjusting the spatial configuration of the chiral center in cross-linker, we can produce peptides with helical or non-helical structures while possesses the same chemical compositions. This discovery not only establishes a new methodology for constructing stapled peptides but also provides a possible explanation for the mechanism of peptide helical structure generation.

Focusing on this important discovery and its biological application, Dr. Kuan Hu completed a series of related researches and published a series of articles in many decent journals. These works span many fields such as chemical methodology, cell biology, computational chemistry, and cancer biology, and constitute the main content of this thesis.

During the five years as a Ph.D. candidate in my group, Dr. Kuan Hu, with his consistent steadfastness and diligence, left me with a deep impression. For his future career, the doctorate is just a starting point, and he will confront many difficulties and challenges afterward. Here, I wish that he will be able to overcome obstacles and make achievements in his future scientific research road.

Shenzhen, China

Prof. Zigang Li

# Preface

Regulation of aberrant intracellular Protein-Protein Interactions (PPIs) is a promising strategy for disease management. Constrained peptides that are fixed into a helical conformation by chemical means are a representative molecular modality to modulate PPIs. Many peptide stabilizing strategies have been established and been widely used in the past decade. However, the correlation between peptides' secondary structure and their biophysical and biochemical properties remains elusive. The core question is how the secondary structure influences a peptide's cell permeability, binding affinity, and serum stability. To answer this question, the probable biggest obstacle is a lack of a method to obtain peptide epimers that possess the same chemical composition but distinct secondary structures.

To overcome this limitation, I designed a novel peptide stapling strategy, named as "chiral center induced peptide helicity (CIH)", for which the major feature is the introduction of a precise chiral center in the peptide crosslinker. I found that this precisely positioned carbon chiral center determines the secondary structure of a peptide, and the *R*-configured chiral center induces a helical conformation while the **S**-configured chiral center results in a random coil. Moreover, I proved that the peptides with *R*-configuration chiral centers display enhanced cell permeabilities and target binding affinities than the S-configuration chiral center counterpart peptides. These results unambiguously demonstrate that helical conformation is good for the pharmacological properties of stapled peptides. This CIH method also provides a robust platform to investigate the relationship between peptide secondary structure and bioactivities.

Harnessing the CIH method, I then developed peptide inhibitors targeting the p53 and MDM2/MDMX interactions, which is a typical PPI and has been demonstrated a potential target for cancer intervention. I first designed a series of CIH peptides with different in-tether substitution groups or peptide sequences, among which two peptides, namely MeR and PhR, effectively restored the function of p53. The restoration of p53 function leads to cell proliferation inhibition and apoptosis induction in multiple p53-wild type cancer cells, for instance, breast cancer cell line MCF-7 and ovarian teratocarcinoma (PA-1) cancer cell line. Moreover, these peptides showed little toxicity towards normal cells or cancer cell lines with mutated p53. The *in vivo* study of the peptides in the PA-1 xenograft model showed a tumor growth rate inhibition of 70% with a dosage of 10 mg/Kg (one injection every other day). As far as we know, the PhR is the first peptide inhibitor that targets a tumor with stemness. This study will open a new avenue for cancer stem cell therapy in the future.

Shenzhen, China

Kuan Hu

# Parts of this thesis have been published in the following journal articles:

[1] **Hu, K.**, Geng, H., Zhang, Q., Liu, Q., Xie, M., Sun, C., Li, W., Lin, H., Jiang, F., Wang, T., Wu, Y.-D., and Li, Z. (2016) An In-tether Chiral Center Modulates the Helicity, Cell Permeability, and Target Binding Affinity of a Peptide, *Angewandte Chemie International Edition* 55, 8013-8017.

[2] **Hu, K.**, Li, W., Yu, M., Sun, C., and Li, Z. (2016) Investigation of Cellular Uptakes of the In-Tether Chiral-Center-Induced Helical Pentapeptides, *Bioconjugate Chemistry* 27, 2824-2827.

[3] **Hu, K.**, Sun, C., and Li, Z. (2017) Reversible and Versatile On-Tether Modification of Chiral-Center-Induced Helical Peptides, *Bioconjugate Chemistry* 28, 2001-2007.

[4] **Hu, K.**, Yin, F., Yu, M., Sun, C., Li, J., Liang, Y., Li, W., Xie, M., Lao, Y., Liang, W., and Li, Z.-g. (2017) In-Tether Chiral Center Induced Helical Peptide Modulators Target p53-MDM2/MDMX and Inhibit Tumor Growth in Stem-Like Cancer Cell, *Theranostics* 7, 4566-4576.

[5] **Hu, K.**, Sun, C., Yu, M., Li, W., Lin, H., Guo, J., Jiang, Y., Lei, C., and Li, Z. (2017) Dual In-Tether Chiral Centers Modulate Peptide Helicity, *Bioconjugate Chemistry* 28, 1537-1543.

[6] **Hu, K.**, Sun, C., Yang, D., Wu, Y., Shi, C., Chen, L., Liao, T., Guo, J., Liu, Y., and Li, Z. (2017) A precisely positioned chiral center in an i, i + 7 tether modulates the helicity of the backbone peptide, *Chemical Communications* 53, 6728-6731.

# Acknowledgements

At the very beginning, I would like to express my sincere thanks to my supervisor, Prof. Zigang Li, for the continuous guidance and encouragement during the entire period of my study. I joined Prof. Li's group in the year of 2012, and I feel so grateful and honored for his excellent guidance in both research and life. His dedicatory attitude to research, as well as the patient and fairy attitude to students, has taught me a great lesson in my life. The strict training I received, the valuable experience I gained, and especially the attitude to work I learned from him would no doubt become the most precious golden mine in my future life.

I would like also to thank Prof. Yundong Wu, Prof. Tao Wang, Prof. Xinwei Wang, and Prof. Yongjuan Zhao for the kind help and precious suggestions they provided during the collaboration. Their insistent attitude to research, insightful thoughts and equitable and mutually respectful dialogue have greatly encouraged and inspired me.

My gratitude is also given to my dear group members and friends for their sweat accompanying in life and helpful discussion in research. I shall always keep in mind the happy and meaningful memories we spent together in the past years. They are, Dr. Hao Geng, M. S. Chengjie Sun, Dr. Feng Yin, Dr. Huacan Lin, M. S. Mengyin Yu, Dr. Yixiang Jiang, Dr. Xiaodong Shi, Dr. Qisong Liu, Dr. Bingchuan Zhao, M. S. Miao An, Dr. Yuan Tian, Dr. Qingzhou Zhang, and M. S. Chengxiang Lei, and so on.

Thanks are also extended to all the staff members in the School of Chemical Biology and Biotechnology, also the Peking University Shenzhen Graduate School for the academic atmosphere and the service they offered.

Finally but most importantly, with all my heart, I would thank my parents for their endless love and support from my first day to the end of the days. I would also thank my girlfriend Hui Wang for warm companying all the time.

# Contents

# Chapter 1
# Introduction

## 1.1 Introduction of Protein-Protein Interactions

### *1.1.1 Significance of Studying Protein-Protein Interactions*

Protein-protein interactions (PPIs) play important roles in cell life activities, such as signal transduction, cell cycle, immune system, gene regulation, and so on [1–3]. The size of protein-protein interactions is one of the important indicators to measure the complexity of life. Humans contain over 650,000 pairs of PPIs. The study of PPIs is of great significance, these benefits include but not limited to:

1. To help better understand system biology;
2. To provide novel strategies for disease diagnosis and treatment;
3. To aid the development of new drugs.

The development of PPIs targeting pharmaceuticals is significant, however, PPIs are traditionally considered undruggable by small-molecular drugs. Despite small-molecular drugs that can penetrate cell membranes readily, their specific binding to PPIs is a huge challenge and is currently an important issue in the chemical biology community. Unlike active catalytic pockets displayed in enzymes, the interface of protein-protein interaction is large as well as flat and lacks obvious binding pockets. To design effective PPI inhibitors, it requires molecules to meet the following conditions: providing continuous non-covalent interactions with target proteins; the contact area with protein is greater than 800 $Å^2$, a value significantly larger than the size of a traditional substrate-enzyme binding pocket. Notably, although the contact area of PPIs is relatively large, the key interactions are usually dominated by a few of key amino acids, which are designated as 'hot spots' [4, 5]. This truth makes the rational design of PPI inhibitors achievable and possible. Indeed, the current rationale in designing PPI inhibitors is to synthesize peptide mimetics that reproduce the hot



K. Hu, *Development of In-Tether Carbon Chiral Center-Induced Helical Peptide*, Springer Theses,
https://doi.org/10.1007/978-981-33-6613-8_1

spots aligned in the surface of specific PPIs. Based on this, a lot of promising PPI inhibitors have been developed for a variety of targets [5–7].

### 1.1.2 Classification of Protein-Protein Interactions

In general, the area of the PPIs interaction interface is between 1000 and 6000 $Å^2$ [8]. When the interaction area is less than 2000 $Å^2$, the entire interaction position is limited to one site. For PPIs with a larger interaction area, their interaction area is composed of several scattered areas, and these distributed areas are separated by amino acids exposed to the solvent [9]. It is worth pointing out that there is no simple linear correlation between the interaction area and the binding ability. Compared with non-critical sites, hot spot amino acids play a key role in binding [4]. A data analysis based on an alanine scanning experiment shows that tryptophan (Trp), tyrosine (Tyr), and arginine (Arg) usually take the roles as hot amino acids. Besides, polar aspartic acid (Asn) and histidine (His) have high opportunities as hot amino acids (Fig. 1.1).

PPIs can be interactions between homologous proteins or heterologous proteins. According to the strength of the interaction, it can be divided into forced (strong effect and persistent) and non-forced (weak and temporary) [1]. The difference in affinity between different PPIs can vary by six orders of magnitude, from picomoles to moles. Under normal circumstances, the interaction interface is considered to be hydrophobic, which is surrounded by a ring composed of polar amino acids [10], or is composed of a mixed hydrophobic region in which polar interactions and water molecules are distributed [11]. The complex of strong protein-protein interaction is similar to a large globulin, and the interaction interface is similar to the internal structure of each globulin. In contrast, the interface of weak protein-protein interactions is usually much smaller and shows an elusive hydrophobic interaction cross-section, which indirectly reflects that the exposure of the hydrophobic region to the solvent is detrimental to binding [12].

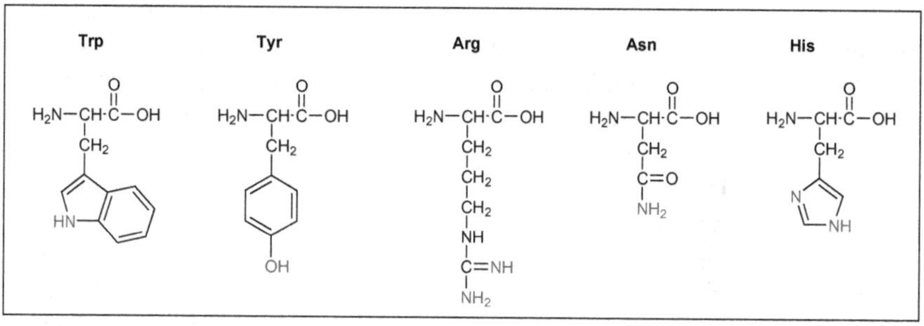

Trp Tyr Arg Asn His

**Fig. 1.1** Amino acids as hot spots in protein-protein interactions

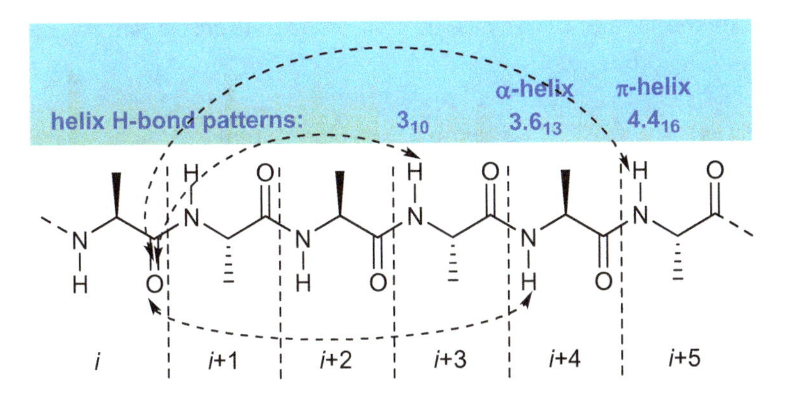


**Fig. 1.2** Types of peptide helix and the corresponding Hydrogen-bond patterns

### *1.1.3 Introduction of Peptide Helix*

Helical structure is the most abundant secondary structure of peptides, which accounts for 30–40% of the protein structures [13]. Helical peptides are stabilized by continuous intramolecular hydrogen bonds (i, i + n). The most common way of naming polypeptide helices is based on the number of amino acids required to form a single helix, and the atoms contained in the macrocycle formed by the hydrogen bond formed by the carbonyl group of the i amino acid and the amino group of the i + n amino acid. The number is defined as a subscript (Fig. 1.2) [14, 15].

In natural proteins, only helix composed of three to five amino acids have been observed ($3_{10}$, $3.6_{13}$, $4.4_{16}$), although other helical forms are theoretically feasible [16]. The $3_{10}$ helices are composed of continuous β corners. The $3.6_{13}$ and $4.4_{16}$ helices consist of successive α and π turns, respectively. The π helix is rarely observed in proteins [17, 18], and the $3_{10}$ helix accounts for about 10% of all protein helical structures [18]. The remaining 90% is α helices. According to the available structural data analysis [19], the helix occupies 62% of the PPIs interface [20], therefore, in this case, the role of α helix is particularly important. A series of methods for mimicking the helical structure of peptides have been reported, and the most used four methods are shown in Fig. 1.3.

A loop of a peptide α helix is composed of 3.6 amino acids, and the length of each helix is 0.54 nm. The dihedral angles θ and φ are −60° and −45°, respectively. According to the spatial arrangement of amino acid side chains, a peptide α helix can be considered to be composed of three faces. In nature, the proportion of helix in the secondary structure of protein exceeds 30%, which is the highest proportion of all peptide secondary structures. Therefore, a considerable part of protein-protein interactions is mediated by α helices, making α helix the most important structural template for designing and regulating PPIs inhibitors. By mimicking a helical sequence involved in the interaction on the interface of PPIs and retaining the key

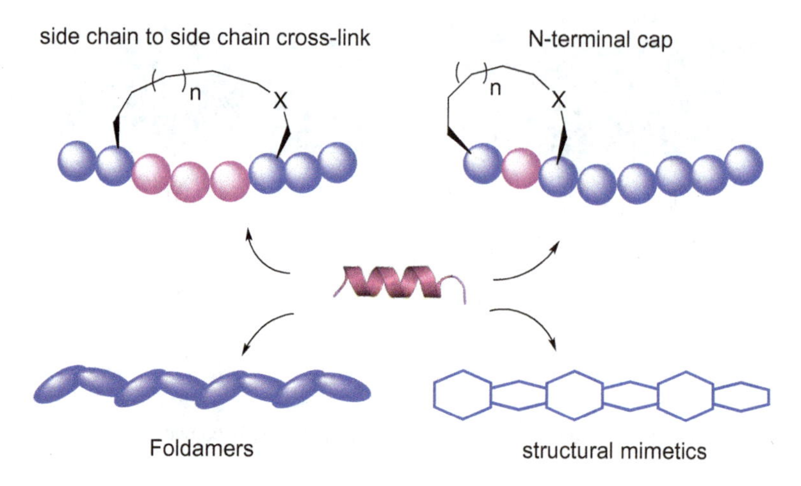


**Fig. 1.3** Methods for constructing helix mimetics

sites involved in the binding, the design of new peptidomimetics is a commonly used method for designing PPIs inhibitors [21].

The types of helix-mediated PPIs are also diverse. PPIs such as the p53/hdm2 interaction is mediated by peptides with a length of 3–4 helices [22, 23], and the key amino acid side chains are distributed in the same face of the peptides. Other interaction modes are considered to be different from those in nature. Taking the glycoprotein gp41 as an example, gp41 hexamers are formed by the formation of trimeric coiled-coil and then stacked by N-terminal and C-terminal [24]. The estrogen receptor firmly binds to the coactivator protein on the surface throughs interaction with two faces of the peptide and a clamp formed by the first and last charged amino acids [25]. Some PPIs are even involved in the three faces of the helix, and the corresponding protein is involved in the binding through a relatively shallow and long slit. The design of peptide inhibitors for these targets is more difficult. According to the screening method called alanine scanning, the Arora group has carried out statistics on the interaction surface of the helix binding to the target proteins (Fig. 1.4) [20]. The physiological functions of proteins are classified according to the number of surfaces involved. At the same time, the Arora group searched the PDB database to identify and characterize a large number of protein interaction interfaces involving helix [20]. Bergey et al. published a database (HiPP), which contains data on the interface involved in helix in protein-protein interactions, which provides a rich data basis for research in this field. HiPP contains information about the length of the interface helix. In the 2013 version, 7308 helical sequences were incorporated. The shortest sequence consists of 4 amino acids, and the average sequence length is 13 amino acids.

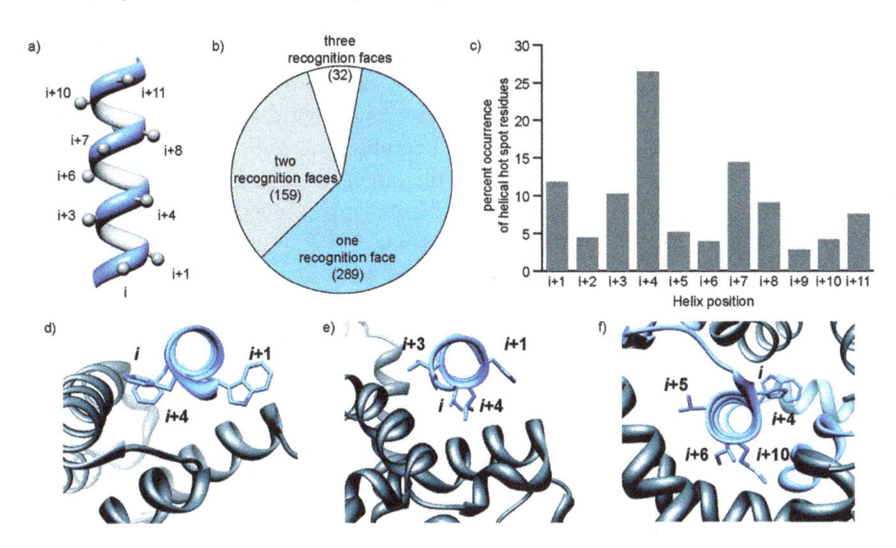


**Fig. 1.4** Classification of peptide-protein binding patterns. (**a**) Energetic contributions of residues on different faces of interfacial helices. (**a**) The positioning of side chain residues on a canonical α-helix. (**b**) Percent occurrence of hot spot residues on one, two, or three helical faces (the total number of helices in each category is shown in parentheses). (**c**) Percent occurrence of hot spot residues as a function of helix position. (**d–f**) examples of protein complexes with hot spot residues on one face, two faces, and three faces. Reprints with permission from ref. 20

## 1.2 History and Current Status of Peptide Drugs

Polypeptides as drugs for disease treatment can be traced back to the 1920s. At that time, people had used insulin purified from bovine pancreas to treat children's smallpox. After that, endogenous peptide molecules were proved to be useful for disease treatment [26]. Initially, the use of polypeptides to treat diseases was a particularly difficult task, which was plagued by deficiencies such as difficulty in synthesis, low yield, difficulty in purification, low stability, and complicated administration methods of peptides. With the development of new technology, these problems are gradually solved. Two milestone breakthroughs in the field are the invention of solid-phase peptide synthesis technology in the 1960s and the gradually matured peptide separation and purification technologies, such as the use of high-performance liquid chromatography (HPLC). These achievements in the field of peptide chemistry make the large-scale synthesis and purification of peptides easier and cheaper. Other the other hand, the methods for screening drug candidates from endogenous peptides are also more diverse [27].

At present, up to 60 peptide drugs have been approved for use in the treatment of human diseases, and there are about 140 peptide molecules in different stages of clinical trials. Most of these molecules are discovered based on natural sequences of endogenous peptides. In 2010, the total annual global sales of the four top peptide drugs exceeded US$1 billion. The market share of peptide drugs is still growing exponentially and is expected to reach US$25.4 billion in 2018. Notably, there are still many challenges for endogenous peptide sequences as drugs. These molecules are usually metabolically unstable and therefore have poor oral availability and membrane penetration. Linear peptides have a short half-life, therefore, the effective drug concentration to reach the target tissue is insufficient. The short half-life in plasma is mainly caused by the enzymatic degradation in the blood, liver, and kidneys, and it is also quickly eliminated by the kidneys. Considering that oral peptide drugs will be digested by the digestive system, many peptide drugs are administered through intraperitoneal and intravenous injection [28, 29].

The existing deficiencies of polypeptide drugs can be modified or chemically altered to improve drug activity and reduce possible side effects. For example, desmopressin (DDAVP) is obtained by modifying an amino acid of vasopressin. This change can reduce the risk of high blood pressure caused by vasopressin. Another method is to design interface peptides that specifically bind to a specific area on the surface of a target protein [30]. The interface peptide strategy has been developed for decades. The core idea of interface peptide is that the critical sequence in the binding interface of a PPI can specifically inhibit the interactions between the two proteins. As shown in Fig. 1.5, it is a general flowchart for designing interface peptide inhibitors [31]. Based on this strategy, several important protein-protein interactions were effectively blocked by their interfaces peptides, for example, Bcl-2/Bax, P53/MDM2.

The development of peptide drugs has received unprecedented attention and achieved extraordinary achievements in the past decade. This phenomenon is largely due to the emergence of a new class of drug modality-stapled peptides. Stapled peptides are structurally stabilized peptides that adapt α-helical conformation enforced by an artificial side chain crosslinker. The stapled peptides are very important in interfering with disease-related protein-protein interactions. Compared with other types of drug modalities, they show remarkable pharmacokinetic advantages, such as high affinity, high specificity, and resistance to protease degradation. A series of stapled peptides have been devised, as either a single-agent treatment method or combination therapies. Stapled peptides are considered a breakthrough with huge potentials and may even change the blueprint of a protein drug development [31–35].

In the past ten years, the paradigm of disease treatment has undergone fundamental changes from a methodological point of view. Specific targeted therapy has become the main direction of disease treatment and dominates the trend of future drug development. Understanding the intracellular processes of drug molecules and the biological molecules that interact with them are the most critical tasks in the development of drugs for specific diseases. Targeted therapy has led to the arrival of a new era of drug development, especially the exponential growth of new drug development strategies for specific cancers. These recently developed drug molecules can be categorized to two types: small-molecular and large-molecular drugs. Small-molecular drugs are a class of organic/inorganic substances that are structurally stable through chemical synthesis. Although most small molecules can effectively penetrate cell membranes and target specific proteins in cells, protein-protein interactions have a large surface area and a shallow binding pocket, making small-molecular drugs

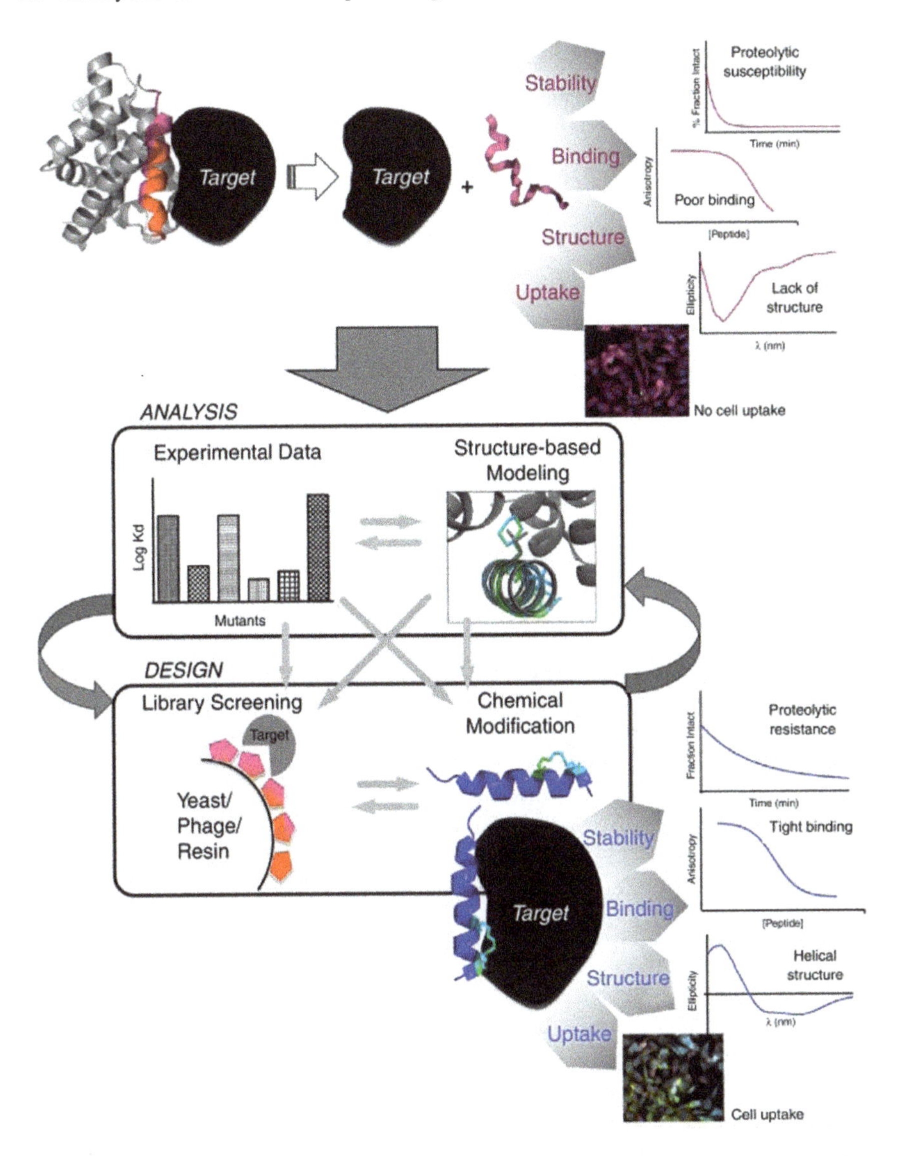


**Fig. 1.5** Overview of the design flow for helical peptide inhibitors of protein-protein interactions. Reprints with permission from ref. 31

are ineffective to these targets [36, 37]. In contrast, large-molecular drugs such as biologicals have many advantages in comparison with small-molecular drugs. They can effectively interfere with extracellular PPIs. However, macromolecular drugs cannot effectively penetrate the plasma membrane of the cell and are more vulnerable to enzymatic degradation. Due to the inherent limitations of the two classes of drug molecules, nearly 80% of the disease-associated protein targets are considered

Table 1.1 Merits and demerits of different molecular weight drugs

| Comparison of different molecular weight drugs | | | |
|---|---|---|---|
| | Small molecules | Biologicals | Peptides |
| Molecular weight | $<10^3$ | $>10^4$ | $10^3 - 5*10^3$ |
| Merits | High stability | Low toxicity | Low toxicity |
| | High permeability | High affinity | Medium affinity and selectivity |
| | Low cost | High selectivity | Outer and inner cellular targets |
| | | Outer cellular targets | Low cost |
| | | | Deep tissue penetration |
| Demerits | Severe side effects | Poor in vivo stability | Poor in vivo stability |
| | Weak ability to target | High cost | Fast pharmacodynamics |
| | membrane receptors | Inability to traverse cell membrane | |

'undruggable.' In this case, stapled peptide, as a complementary drug modality, is emerging as a robust tool to interrogate these difficult targets that are traditionally considered unaddressable by small molecules, which overcomes the limitations of small molecules and macromolecules. They can not only target extracellular PPIs but also target intracellular PPIs owe to their remarkable cell penetration and high stability under physiological conditions (Table 1.1) [38–40]

## 1.3 Methodology for Stabilizing Peptide Secondary Structures

It is necessary to develop modulators that regulate protein-protein interactions. They can not only help explain a lot of life processes but also are potential drug candidates [41]. The protein-protein interaction interfaces are much different from the binding pockets of enzymes or receptors, and they are usually composed of large and shallow groove-like regions. Therefore, small-molecular drugs usually show limited potency in modulating PPIs. This situation motivates us to develop new drug modalities targeting PPIs [33, 42, 43]. The peptide binding epitope in the protein interaction interface serves as the starting point for the design. These epitopes have specific secondary structures. According to the conformation of the peptide, these secondary structures can be divided into β sheet, α helix [16, 44], and turn structures [45, 46]. The problem is that if the interface peptide is departed from the parent structure with a stable environment, the regular secondary structure will be lost. When in solution, the interface peptides' structure exchanges among all possible secondary structures. This flexible structure makes the interface peptide easy to be degraded by the enzymes, and at the same time, it is not conducive to the binding of the target protein because of great entropy consumption during binding [47, 48].

Furthermore, most of the short linear peptides lack sufficient membrane penetration ability. A large number of researches suggested that helical conformation could help peptide cross the cell membrane. To increase the cell penetration of peptides, methods for enforcing peptides in helical conformation have been established. Introducing amino acids that easily form $\alpha$ helix, such as Aib [49], and covalently coupling the side chains of amino acids at positions i, i + 4, and i + 7 in peptides, are commonly used methods to enhance peptide helicity. The early reported cross-coupling methods include disulfide bond formed by two cysteines [50], or amide bridge formed by lysine and glutamic acid or aspartic acid [51]. Date back to the beginning of this century, a new method named all-hydrocarbon stapled peptides that were yielded by olefin metathesis reaction is reported by Professor Verdine et al. Compared to constrained peptides generated by other cross-coupling methods, the all-hydrocarbon stapled peptides exhibit higher structural rigidity. Moreover, the hydrophobic crosslinker makes the peptides greater cell permeability [52].

In addition to ruthenium-catalyzed metathesis, other chemical bioconjugation methods for constructing constrained peptides have also been reported. For example, the bis(alkylation) of thiol-containing peptides [53–56]. Indeed, cysteine-based cyclization has become the most important strategy for constructing cyclized peptides. This method has many advantages. The peptide sequences containing designated cysteines can be synthesized via the recombinant method, which enables large library screening of highly potent peptide inhibitors for protein targets. Moreover, the cysteine-based cyclization can be occurred in ambient conditions to avoid the use of any toxic heavy metal catalyst.

Recent papers indicate that new chemical techniques can construct side chain 'tethered' peptides on the surface of phages. In a recently published paper, Wang et al. demonstrated that olefin-sulfhydryl coupling can be used to construct helical peptides, which can selectively block the P53/MDM2 interactions and further induce apoptosis of P53-wild type cancer cells [57]. Pentelute et al. used perfluorobenzene as a linking group to synthesize a constrained peptide inhibitor of HIV-1 capsid assembly polyprotein [58, 59]. The side chain tether constructed by perfluorinated groups is lipophilic and can improve cellular uptake. Muppidi et al. used a series of aryl and vinyl aryl groups with matching lengths to construct macrocyclic peptides and studied the relationship between membrane penetration and the linker types [56].

Compared to the 'inert' crosslinker without active functional groups, a modifiable tether possessing one or more modification sites are more attractive for synthesizing multifunctional peptide inhibitors. Based on this, a series of modifiable linkers were employed in stapled peptides. The use of modifiable crosslinkers provides possibilities to further optimize the peptides' bioactivity via a second modification. For example, Spring et al. developed an i, i + 7 double click stapling method, and based on this method they synthesized a series of constrained peptides that bind to the MDM2 protein, and demonstrated the binding ability of these peptides to MDM2 protein is at nanomolar [60]. However, the peptide with the highest binding affinity has a weaker cell penetration ability, and the intracellular p53 reporter gene experiment shows that the peptide is almost ineffective in vivo. The possible reason for this phenomenon is that negatively charged amino acids in the peptide interfere with the binding to the phospholipid bilayer membrane, which prevents the peptide from penetrating the cell membrane [61, 62]. In our study, we developed peptide constraining methods based on the thiol-ene photoreaction. The thiol-ther crosslinker generated by this method can be further modified at the sulfur atom, such as sulfonium modification. As a result of this modification, the cell-penetrating ability of the peptide is increased because an excess positive charge was added to the peptide. We also demonstrated the sulfonium modification can modulate the binding affinity of the peptide [63–66]. Among other modifiable crosslinking methods, Vasco et al. proved that the Ugi reaction can be used to construct α-helix peptides, and after ring formation, exocyclic functionalities as N-substituents were incorporated [67].

In another example, Assem et al. reported a macrocyclization strategy using dichloroacetone as the linking group. When this group reacts with the nucleophilic thiol group on the peptide, the helical conformation of the formed cyclic peptide is remarkably stabilized. Aside from stabilizing helical structures, the ketone moiety embedded in the linker can be modified with diverse molecular tags by oxime ligation. The reversible properties of the oximes make it useful for the dynamic covalent chemical modification of peptides, which will help improve the selectivity of the peptide and optimize the interaction with the target [68].

A switchable linking group achieves the effect of controlling the structure of peptides by using light or other conditions to change the conformation of the linking group. The use of cysteine to couple with azobenzene-containing linking groups has been demonstrated to be an effective photoswitchable strategy for light-regulating the physiological activity of peptides [69]. Martin-Quiros et al. used this method to construct a photoswitchable peptide inhibitor of β-arrestin/β-adaptin protein-protein interactions, which has a 12.6-fold difference in binding constant in the presence or absence of light [70]. More interestingly, Belotto et al. proved that peptides containing azobenzene linker can be selectively bound to protein targets of choices in the phage display screening, and the affinity of the peptide could be modulated by UV light. This method is robust and can be applied for the in vitro evolution of photoswitchable ligands to any targets [71]. In another reversible α-helix formation method, Miller et al. developed a reversible hydrogen bond surrogate method that utilizes an internal disulfide linkage. Structural analysis indicates that the dynamic nature of the disulfide bridge allows for the reversible formation of an α-helix through oxidation and reduction reactions [72]. The most commonly used methods for stabilizing peptide α-helix are introduced below.

1. Side chain-side chain coupling for helix stabilization

Peptide helix forms intramolecular hydrogen bonds from the carbonyl oxygen of the i-th amino acid and the protons of the amino group of the i + 4th amino acid. Other stabilizing effects can be produced by forming a salt bridge between two amino acids (such as glutamic acid and lysine) on the same face. This type of stabilization was used to stabilize the helical structure of peptides in early research [73]. Thereafter, in the peptide, a covalent bridge linker is formed by either the i, i + 3 amino acids or i, i + 4 amino acids to stabilize a single helix, or a covalent bridge is formed in the i, i + 7 amino acids to stabilize two helical turns. Early cyclization methods included the formation of amide bonds between glutamic acid and lysine residues [51], or the formation of disulfide bonds [74]. Then the combination of the side chain coupling strategy and the helix promotion effect of α carbon methylation [75] led to the production of all-hydrocarbon stapled peptide methodology [76]. These peptides contain all-hydrocarbon crosslinkers formed by the ruthenium-catalyzed metathesis (RCM) reaction of two olefin-containing amino acids. More recently, a protein stapling technology called genetically encoding was reported [77]. In this method, an unnatural amino acid containing an electrophilic side chain is introduced into a specific position in a protein by genetic engineering technology. The amino acid can cross-coupling with adjacent nucleophilic amino acids (lysine, histidine, cysteine) on the protein. This method is designed to stabilize α-helix sequences in the protein. When designing stapled peptides with crosslinked side chains, it is necessary to pay special attention to the position of the bridges and the length of the side rings. Moreover, it should be noticed that a careful selection of a combination of helix promoting factors and avoiding unfavorable side chain interference with the target protein are two basic aspects that should be considered in the design of stapled peptides. The present methods used to construct PPIs inhibitors mainly include coupling methods based on thiol-ether, amide, triazole, and all-hydrocarbon crosslinkers (Fig. 1.6).

1. Crosslinking based on sulfhydryl groups

One of the earliest strategies used to construct structurally constrained helical peptides was the formation of disulfide bonds by two Cysteine amino acids at positions i and i + 7. To ensure the correct orientation of the side ring, one D-type Cysteine was used. Compared with unconstrained peptides, cross-coupled disulfide cyclic peptides show higher helicity [74]. The cross-coupling between the D-cysteine at position i and the L-cysteine at position i + 3 has also been shown to stabilize the α-helical peptide [78]. The length of the side chain was shown to affect the helical content and target recognition. Replacing cysteine with homocysteine can also form a stable helical peptide [79, 80]. However, disulfide bonds are easily reduced by the reductive species in eukaryotic cells. Therefore, more chemically stable thioether bonds are used to replace disulfide bonds. Cysteine is an amino acid with specific nucleophilic ability among natural amino acids, so the corresponding electrophilic group can be designed to selectively react with sulfhydryl moieties. A series of amphiphilic linkers are designed to couple with two cysteines in a peptide to form side chains to stabilize the peptide structure. Among different electrophilic molecules, m-xylene reacting with two L-cysteines at positions i, i + 4 was demonstrated the highest helical content proved by circular dichroism and nuclear magnetic resonance [53]. For longer peptides, the helical content of the constrained peptides that are formed by brominated diaryl xylenes reacting with L-cysteine at position i and D-cysteine at position i + 7 is the highest. These side chain-coupled PPI inhibitors showed higher penetration ability compared to none constrained peptides [54–56].

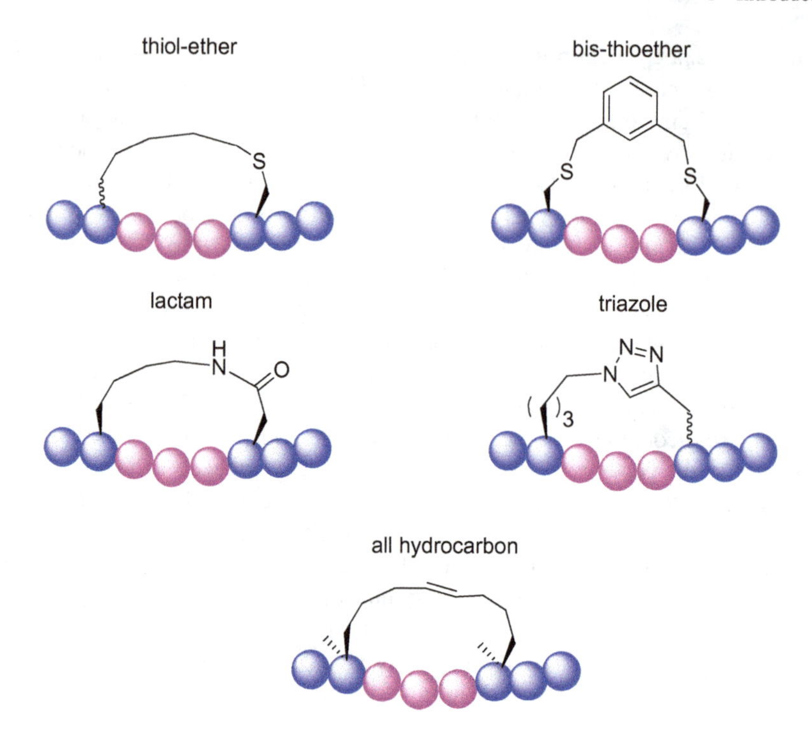


**Fig. 1.6** Stapling chemistries for peptide helix stabilization

Moreover, perfluorobenzene has also been used to construct stable helical peptides. Fluorination was shown to further enhance the anti-degradation ability of peptides as well as improve the membrane penetration ability [58].

2. Amide bond crosslinking method

Contemporaneous with the formation of disulfide bonds, the formation of helical peptides was induced by the formation of amide bonds between lysine (Lys) at position i and aspartic acid (Asp) at position i + 4 [51]. Furthermore, the diamidated linker is also used to form a diamide bond with two glutamic acids at positions i and i + 7 in a peptide to constrain peptides [50, 81]. PPI inhibitors constructed by forming two adjacent amide bonds have also been reported. Of course, two overlapping amide bond crosslinkers are also used to construct high helicity polypeptides [82, 83]. Although helical peptides constructed by amide bonds have high helicity, amide bonds are not preferable to penetrating cell membranes, so the corresponding applications are also limited [83].

3. Triazole crosslinking method

Triazoles are generated by copper-catalyzed azide-alkyne groups through [3 + 2] cycloaddition reactions, which are a very important type of click reactions and play a significant role in organic synthesis and drug development [84]. This reaction can also be used to stabilize the peptide α helix. Studies have shown that using [1–3] triazole to replace the amide ring, the helicity of the obtained peptide is similar to that of the amide ring peptide [85]. In order to obtain a polypeptide with high helicity, two triazole side loops can be introduced into a peptide, and the obtained peptide is more stable and binds more tightly to the target [86]. For peptides containing two azide functional groups located at positions i and i + 7, a linker containing two terminal alkyne groups with a matching length can be used to react with it to form a crosslinker that stabilizes two helices [60, 87]. Based on the triazole linker, Lau et al. developed a modifiable linker that can be conjugated with arginine to improve the cell permeability, which was proved by a reporter gene experiment [61, 88].

4. All-hydrocarbon crosslinking

The all-hydrocarbon stapling method for peptides was invented by Verdine et al. [76]. This method has two main characteristics when stabilizing helical peptides. First, the methylation of the α carbon of the unnatural amino acid of crosslinking. Second, the utilization of olefin metathesis to generate peptides with all-hydrocarbon side chain crosslinker. They first systematically studied the length of the crosslinker, the chirality of stapling residues, and the coupling mode. For crosslinker spanning one peptide helix, they found that a full hydrocarbon crosslinker ($S_5 + S_5$) of 8 atoms with the S5 amino acids in the i and i + 4 positions in the peptide has the highest cyclization efficiency and helix inducing ability. The case for stapling two helices is a crosslinker of 11 atoms (S5 + R8) with the S5 at the i position and the R8 at i + 7 position. In the comparison of different types of crosslinkers, they found that the i, i + 7 side chain system showed the best helix inducing ability. Moreover, the stapling peptides were revealed greatly improved stability to protease degradation. About ten years later, this group further published results of how the chirality of amino acids for stapling affects the membrane permeability of the peptides. They also expanded the all-hydrocarbon system to i, i + 3 side chain systems. They found that when the unnatural amino acids are all L-type, the peptide penetration ability is the best, and the i, i + 3 system is not as good as the i, i + 4 system [89, 90]. However, the expansion of the i, i + 3 system allows them to selectively synthesize bicyclic peptides (double staple) in one step by collocation of amino acids with different chirality, using olefin metathesis reaction [91]. In 2013, they explored the influence of chain length and double bond position on i, i + 3 and i, i + 4 systems, respectively. In 2014, to improve upon the stapling technology for stabilizing a peptide in a bioactive α-helical conformation, they report the discovery of an efficient and selective bis ring-closing metathesis reaction leading to peptides bearing multiple contiguous staples connected by a central spiro ring junction [92]. Compared to stapled peptides, the stitched α-helical peptides showed even increased thermal stabilities, dramatically enhanced stability against chemical denaturation and proteolytic digestion. Moreover, the stitched peptides show greatly improved cell-penetrating ability compared to stapled peptides. These features enable stitched peptides potential ligands for chemical genetics applications, next-generation therapeutic agents, and tools for macromolecule cargo delivery.

The stapled peptides have emerged as a new molecular modality for inhibiting a varieties of disease-relevant intracellular or extracellular protein-protein interactions in vivo. Similar to macromolecules, such as antibodies, the stapled peptides bind to targets with a relatively large contact area, making the stapled peptides with good specificity. Moreover, the stapled peptides can penetrate cell membranes to targets intracellular PPIs that are usually unable for macromolecules [93, 94]. Because of these attractive points, the all-hydrocarbon stapled peptide has therefore been successfully and widely used in modulating many important intracellular protein-protein interactions.

2. N-terminal nucleation template-induced helical peptide

In addition to the side chain coupling strategy, another widely used strategy for stabilizing peptide α-helical structure is the N-terminal nucleation template, in which the introduction of a nucleation template to the N-terminal of the peptides induces the formation of a α-helical conformation of the whole peptide (Fig. 1.7). The molecular basis for this strategy is the helix-coil transition theory of peptides [95–97].

A typical example of this strategy is one that devised by Professor Arora et al., who developed an all-hydrocarbon N-terminal hydrogen bond replacement system (hydrogen bond surrogate, HBS) based on the olefin metathesis reaction, as shown in Fig. 1.8. They replaced the i, i + 4 hydrogen bonds formed by the N-terminal amino acids with covalent carbon-carbon bonds [98–102]. The helix nucleation parameters of the peptides are greatly improved. The HBS peptides can effectively maintain the helical conformation in a solution environment. This method has also been applied to the synthesis of a variety of intracellular protein-protein interaction peptide inhibitors, including peptide modulators for HIF-1α, Ras, and p53/MDM2 [103–105].

The introduction of N-terminal nucleation template to induce peptide α-helix has some advantages in comparison with the side chain coupling strategies. For example, the N-terminal nucleation template can avoid the loss of side chain amino acid residues that are involved in cyclization in side chain crosslinking strategies,

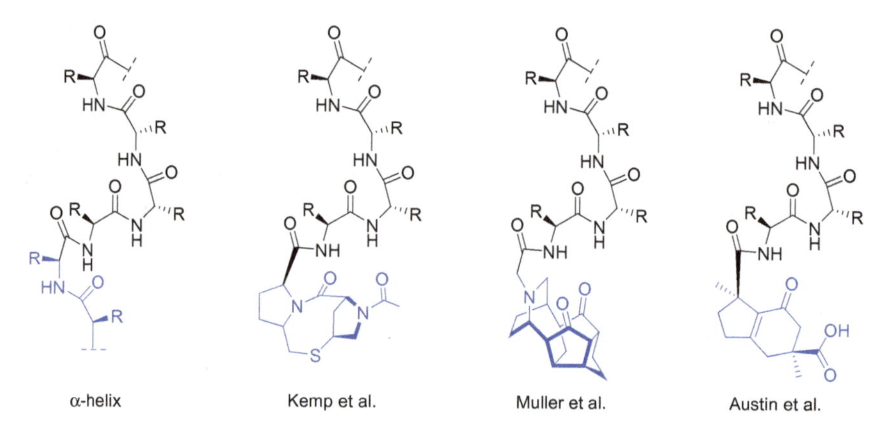


**Fig. 1.7** Examples for N-terminal nucleation template-induced helical peptides

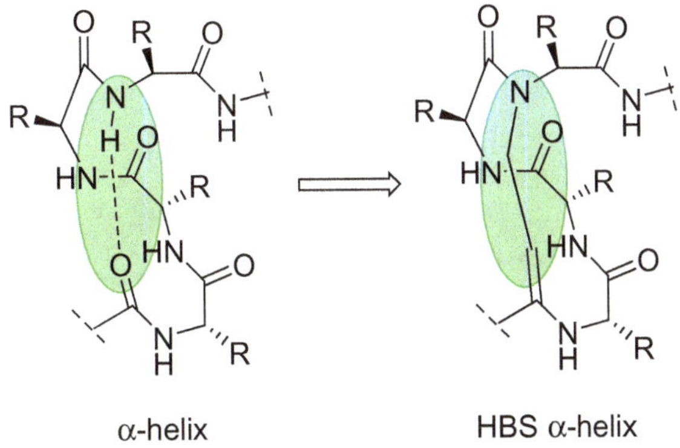

**Fig. 1.8** Illustration of hydrogen bond surrogate system

α-helix HBS α-helix

thus it could prevent any unpredictable interruption of the molecular recognition of the peptides to target, especially for α-helical peptides that bind to targets with multiple faces and even embeds in the target binding grooves. Therefore, the N-terminal nucleation template strategy and the side chain coupling strategy are also complementary to each other in developing PPI inhibitors.

## 1.4 Application of Peptide Stabilization Methodology in the Design of Protein-Protein Interaction Inhibitors

Stable helical peptides have achieved great progress in targeting protein-protein interactions. Among various peptide stapling strategies, the all-hydrocarbon side chain (stapled peptide) is the most widely used method for constructing PPI inhibitors [34, 42, 106]. The diseases involved in these studies include cancer, infectious diseases, metabolic diseases, and diseases in the neural center system (Fig. 1.9). The types of targets directly related to these diseases include transcription factors, receptors, and enzymes [107–127]. According to the position of action, it can be divided into intracellular and extracellular targets. The following chart summarizes the targets of stapled peptides in representative literature. Among these targets, BCL-2 and MDM2 family proteins have been studied the most. Among them, the peptide drug ALRN-6942 targeting MDM2 protein has now entered phase 3 clinical trials. It can be seen that if the drug is proven to have significant efficacy in the treatment of certain cancers, it will greatly promote the development of stapled peptide drugs. Recent research on stable helix peptides is often linked to the hottest fields and applied to viral infectious diseases such as immunotherapy, Ebola, and Zika virus. It is believed that the application of stable peptides will be further expanded.

As there are more and more researches that are investing in stapled peptides, an unambiguous elucidation of the relationship between the peptide structure and the peptide bioactivity is especially important for designing highly potent stapled peptide-based drugs. Scientists have established a comprehensive evaluating system

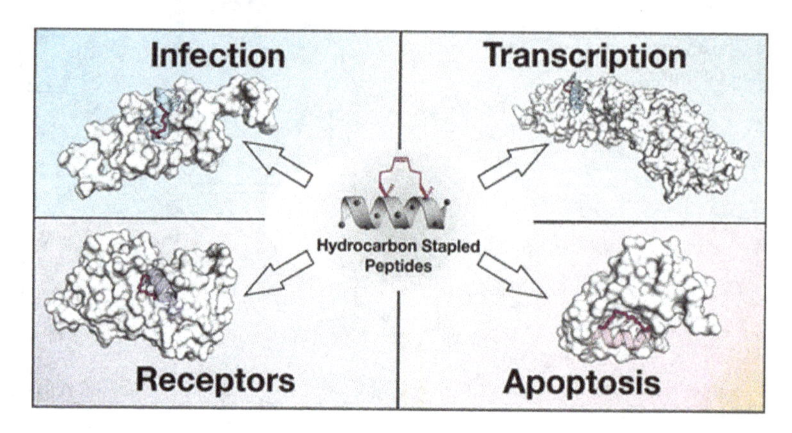


**Fig. 1.9** Applications of hydrocarbon stapled peptides. Reprints with permission from ref. 38

for chemically stabilized α-helical peptides based on the methods for evaluating the activities of small-molecular drugs, these methods span from biochemical assays to cell-based assays. The basic steps include peptide synthesis, lead peptide screening, and target validation, followed by membrane penetration, cytotoxicity, and biological activity test of the peptides by using cell-based assays. Finally, testing the activity of the peptides in animal models. Based on this, a complete evaluation cycle for stabilized α-helical peptide could include the following processes: peptide design and synthesis, solubility study, secondary structure analysis, stability study, target binding assays, cellular uptake, in vitro and in vivo bioactivity, and clinical translation at last (Fig. 1.10) [34].

## 1.5 The Topic Selection Ideas and Research Focus of This Thesis

Based on the above description, it is obvious that the use of chemical methods to stabilize the helical structure of peptides and the application of stapled peptides to combat disease have achieved great success during the past decades. However, despite the knowledge of how and why a crosslinker stabilizes the peptide structure has been well-established, several key questions in this field remain unclear and need to be elucidated, especially the most fundamental aspects regarding the linker-activity relationship. Nowadays, more and more methods were integrated for stabilizing the helical structure of peptides, and for each method, which generates a kind of peptides with distinguished crosslinker features. However, how do these tethers affect the biological properties of the peptides has not been elucidated, although many previous studies have covered this issue. One possible reason is that the existing methods cannot obtain peptide isomers with identical chemical composition but remarkably different secondary structures. This shortage largely damper the investigation of how

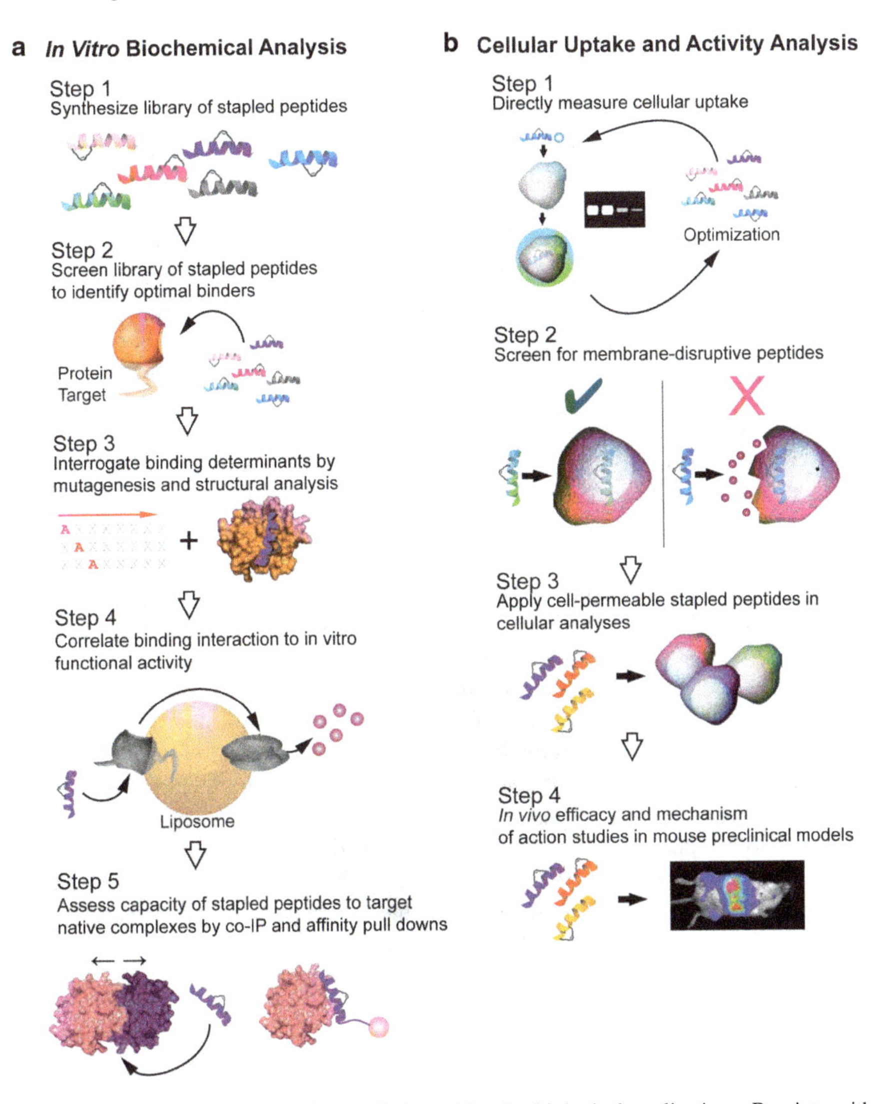


**Fig. 1.10** Workflows for deploying stapled peptides for biological applications. Reprints with permission from ref. 34

the sole structure factor affects the peptides' biological activities. A proper platform that can generate peptide isomers with identical chemical composition but different chemical structures are expected to provide a solution for tackling the dilemma and answer the question of how does the helical conformation influences the peptide activity, such as binding affinity, cell permeability, and in vivo stability.

In the first chapter, the diversity and heterogeneity of protein-protein interactions are introduced. To design effective stapled peptide inhibitors for different targets, there are higher requirements for synthetic methods that can access abundant linker types to expand the chemical space. How to achieve the chemical diversity of helical peptides, obtain more peptides with variable linker structures based on a given sequence, and expand the capacity of the screening library are very important to optimize the binding affinity and finally result in the most promising peptide leads. To tackle the challenge in stapled peptides, this thesis aims to establish several peptide stapling methodology based on theoretical research.

In the second chapter, I introduce the development of a methodology of 'chirality-induced helicity' from the point of view of tether chirality. This system is the first report to introduce a chiral center on the crosslinker of a stabilized peptide. Since this system can obtain peptide isomers with the same chemical composition but completely different structures, it is an ideal platform for studying the relationship between the secondary structure of peptides and their physiochemical properties. In this chapter, I elaborated on the establishment process of this methodology. Then, based on this platform, the relationship between the secondary structure of the peptides and the cell permeability of peptide isomers, the peptides' binding affinity to target proteins, and the biological stability of the peptides were systematically compared.

In the third chapter, we utilized the 'CIH' strategy in developing peptide modulators for P53/MDM2 interactions. The membrane penetration of peptides is one of the key factors that affect the biological activity of peptides in the case of targeting intracellular targets. It is necessary to study the factors affecting the membrane penetration of peptides, especially the influence of the helical structure of peptides on membrane penetration. Because of the remarkable different cell penetration of CIH isomers, it provides a good platform to unambiguously investigate the relationship between bioavailability and secondary structures. The PPI of p53 and MDM2/MDMX, which is an important cancer treatment target, was selected as a model to test the biological activity of the CIH peptide. Both in vitro and in vivo experiments revealed that the P53 mimetic peptide inhibitors stabilized by the CIH method have promising biological activities and safety. Through this study, I demonstrated that the CIH platform is potential for constructing peptide inhibitors for a variety of PPI targets.

## References


1. Jones S, Thornton JM (1996) Principles of protein-protein interactions. Proc Natl Acad Sci U S A 93:13–20
2. Wells JA, McClendon CL (2007) Reaching for high-hanging fruit in drug discovery at protein-protein interfaces. Nat 450:1001–1009
3. Higueruelo AP, Jubb H, Blundell TL (2013) Protein-protein interactions as druggable targets: recent technological advances. Curr Opin Pharmacol 13:791–796
4. Bogan AA, Thorn KS (1998) Anatomy of hot spots in protein interfaces. J Mol Biol 280:1–9
5. Wójcik P, Berlicki Ł (2016) Peptide-based inhibitors of protein–protein interactions. Bioorg Med Chem Lett 26:707–713
6. Arkin MR, Tang Y, Wells JA (2014) Small-molecule inhibitors of protein-protein interactions: progressing toward the reality. Chem Biol 21:1102–1114

7. Azzarito V, Long K, Murphy NS, Wilson AJ (2013) Inhibition of [alpha]-helix-mediated protein-protein interactions using designed molecules. Nat Chem 5:161–173
8. Smith MC, Gestwicki JE (2012) Features of protein-protein interactions that translate into potent inhibitors: topology, surface area and affinity. Expert Rev Mol Med 14:e16
9. Chakrabarti P, Janin J (2002) Dissecting protein-protein recognition sites. Proteins 47:334–343
10. Miller S (1989) The structure of interfaces between subunits of dimeric and tetrameric proteins. Protein Eng 3:77–83
11. Larsen TA, Olson AJ, Goodsell DS (1998) Morphology of protein–protein interfaces. Struct 6:421–427
12. Keskin O, Gursoy A, Ma B, Nussinov R (2008) Principles of protein – protein Interactions: what are the preferred ways For proteins to interact? Chem Rev 108:1225–1244
13. Koch O, Cole J, Block P, Klebe G (2009) Secbase: database module to retrieve secondary structure elements with ligand binding motifs. J Chem Inf Model 49:2388–2402
14. Bragg L, Kendrew JC, Perutz MF (1950) Polypeptide chain configurations in crystalline proteins. Proceedings of the royal society of London. Ser A Math Phys Sci 203: 321–357
15. Tonlolo C, Benedetti E (1991) The polypeptide 310-helix. Trends Biochem Sci 16:350–353
16. Pauling L, Corey RB, Branson HR (1951) The structure of proteins: two hydrogen-bonded helical configurations of the polypeptide chain. Proc Natl Acad Sci 37:205–211
17. Fodje MN, Al-Karadaghi S (2002) Occurrence, conformational features and amino acid propensities for the π-helix. Protein Eng Des Sel 15:353–358
18. Barlow DJ, Thornton JM (1988) Helix geometry in proteins. J Mol Biol 201:601–619
19. Berman HM, Westbrook J, Feng Z, Gilliland G, Bhat TN, Weissig H, Shindyalov IN, Bourne PE (2000) the protein data bank. Nucleic Acids Res 28:235–242
20. Bullock BN, Jochim AL, Arora PS (2011) Assessing helical protein interfaces for inhibitor design. J Am Chem Soc 133:14220–14223
21. Edwards TA, Wilson AJ (2011) Helix-mediated protein–protein interactions as targets for intervention using foldamers. Amino Acids 41:743–754
22. Liu X, Dai S, Zhu Y, Marrack P, Kappler JW (2003) The structure of a Bcl-xL/Bim fragment complex: implications for Bim function. Immun 19:341–352
23. Czabotar PE, Lee EF, van Delft MF, Day CL, Smith BJ, Huang DCS, Fairlie WD, Hinds MG, Colman PM (2007) Structural insights into the degradation of Mcl-1 induced by BH3 domains. Proc Natl Acad Sci 104:6217–6222
24. Chan DC, Fass D, Berger JM, Kim PS (1997) Core structure of gp41 from the HIV envelope glycoprotein. Cell 89:263–273
25. Bruning JB, Parent AA, Gil G, Zhao M, Nowak J, Pace MC, Smith CL, Afonine PV, Adams PD, Katzenellenbogen JA, Nettles KW (2010) Coupling of receptor conformation and ligand orientation determine graded activity. Nat Chem Biol 6:837–843
26. Banting FG, Best CH, Collip JB, Campbell WR, Fletcher AA (1922) Pancreatic extracts in the treatment of diabetes mellitus. Can Med Assoc J 12:141–146
27. White CJ, Yudin AK (2011) Contemporary strategies for peptide macrocyclization. Nat Chem 3:509–524
28. Kaspar AA, Reichert JM (2013) Future directions for peptide therapeutics development. Drug Discov Today 18:807–817
29. Qvit N, Rubin SJS, Urban TJ, Mochly-Rosen D, Gross ER (2017) Peptidomimetic therapeutics: scientific approaches and opportunities. Drug Discov Today 22:454–462
30. Pelay-Gimeno M, Glas A, Koch O, Grossmann TN (2015) Structure-based design of inhibitors of protein–protein interactions: mimicking peptide binding epitopes, Angewandte Chemie (International Ed. in English), 54:8896–8927
31. Rezaei Araghi R, Keating AE (2016) Designing helical peptide inhibitors of protein-protein interactions. Curr Opin Struct Biol 39:27–38
32. Henchey LK, Jochim AL, Arora PS (2008) Contemporary strategies for the stabilization of peptides in the alpha-helical conformation. Curr Opin Chem Biol 12:692–697

33. Hill TA, Shepherd NE, Diness F, Fairlie DP (2014) Constraining cyclic peptides to mimic protein structure motifs. Angew Chem Int Ed 53:13020–13041
34. Walensky LD, Bird GH (2014) Hydrocarbon-stapled peptides: principles, practice, and progress. J Med Chem 57:6275–6288
35. Lau YH, De Andrade P, Wu YT, Spring DR (2015) Peptide stapling techniques based on different macrocyclisation chemistries. Chem Soc Rev 44:91–102
36. Nero TL, Morton CJ, Holien JK, Wielens J, Parker MW (2014) Oncogenic protein interfaces: small molecules, big challenges. Nat Rev Cancer 14:248–262
37. Góngora-Benítez M, Tulla-Puche J, Albericio F (2014) Multifaceted roles of disulfide bonds. peptides as therapeutics. Chem Rev 114:901–926
38. Cromm PM, Spiegel J, Grossmann TN (2015) Hydrocarbon stapled peptides as modulators of biological function. ACS Chem Biol 10:1362–1375
39. Tan YS, Lane DP, Verma CS (2016) Stapled peptide design: principles and roles of computation. Drug Discov, Today
40. Xie X, Gao L, Shull AY, Teng Y (2016) Stapled peptides: providing the best of both worlds in drug development. Future Med Chem 8:1969–1980
41. Milroy L-G, Grossmann TN, Hennig S, Brunsveld L, Ottmann C (2014) Modulators of protein-protein interactions. Chem Rev 114:4695–4748
42. Nevola L, Giralt E (2015) Modulating protein-protein interactions: the potential of peptides. Chem Commun 51:3302–3315
43. Sperandio O, Reynès CH, Camproux A-C, Villoutreix BO (2010) Rationalizing the chemical space of protein–protein interaction inhibitors. Drug Discov Today 15:220–229
44. Ramachandran GN, Sasisekharan V (1968) Conformation of polypeptides and proteins**The literature survey for this review was completed in September 1967, with the journals which were then available in Madras and the preprinta which the authors had received.††By the authors' request, the publishers have left certain matters of usage and spelling in the form in which they wrote them. Adv Protein Chem 23:283–437
45. Chou PY, Fasman GD (1977) β-turns in proteins. J Mol Biol 115:135–175
46. Koch O, Klebe G (2009) Turns revisited: a uniform and comprehensive classification of normal, open, and reverse turn families minimizing unassigned random chain portions. Proteins: Struct, Funct, Bioinform 74:353–367
47. Pelay-Gimeno M, Glas A, Koch O, Grossmann TN (2015) Structure-based design of inhibitors of protein-protein interactions: mimicking peptide binding epitopes. Angew Chem Int Ed Engl 54:8896–8927
48. Sozzani P, Comotti A, Bracco S, Simonutti R (2004) A family of supramolecular frameworks of polyconjugated molecules hosted in aromatic nanochannels. Angew Chem Int Ed 43:2792–2797
49. Banerjee R, Basu G, Chene P, Roy S (2002) Aib-based peptide backbone as scaffolds for helical peptide mimics. J Pept Res 60:88–94
50. Phelan JC, Skelton NJ, Braisted AC, McDowell RS (1997) A general method for constraining short peptides to an α-helical conformation. J Am Chem Soc 119:455–460
51. Chorev M, Roubini E, McKee RL, Gibbons SW, Goldman ME, Caulfield MP, Rosenblatt M (1991) Cyclic parathyroid hormone related protein antagonists: lysine 13 to aspartic acid 17 [i to (i + 4)] side chain to side chain lactamization. Biochem 30:5968–5974
52. Kutchukian PS, Yang JS, Verdine GL, Shakhnovich EI (2009) All-atom model for stabilization of α-helical structure in peptides by hydrocarbon staples. J Am Chem Soc 131:4622–4627
53. Jo H, Meinhardt N, Wu Y, Kulkarni S, Hu X, Low KE, Davies PL, DeGrado WF, Greenbaum DC (2012) Development of α-helical calpain probes by mimicking a natural protein-protein interaction. J Am Chem Soc 134:17704–17713
54. Muppidi A, Wang Z, Li X, Chen J, Lin Q (2011) Achieving cell penetration with distance-matching cysteine cross-linkers: a facile route to cell-permeable peptide dual inhibitors of Mdm2/Mdmx. Chem Commun 47:9396–9398
55. Muppidi A, Doi K, Edwardraja S, Drake EJ, Gulick AM, Wang H-G, Lin Q (2012) Rational design of proteolytically stable, cell-permeable peptide-based selective Mcl-1 inhibitors. J Am Chem Soc 134:14734–14737

56. Muppidi A, Zhang HT, Curreli F, Li N, Debnath AK, Lin Q (2014) Design of antiviral stapled peptides containing a biphenyl cross-linker. Bioorg Med Chem Lett 24:1748–1751
57. Wang Y, Chou DH-C (2015) A thiol–ene coupling approach to native peptide stapling and macrocyclization. Angew Chem Int Ed n/a-n/a
58. Spokoyny AM, Zou YK, Ling JJ, Yu HT, Lin YS, Pentelute BL (2013) A perfluoroaryl-cysteine snar chemistry approach to unprotected peptide stapling. J Am Chem Soc 135:5946–5949
59. Vinogradov AA, Choo ZN, Totaro KA, Pentelute BL (2016) Macrocyclization of unprotected peptide isocyanates. Org Lett 18:1226–1229
60. Lau YH, de Andrade P, McKenzie GJ, Venkitaraman AR, Spring DR (2014) Linear aliphatic dialkynes as alternative linkers for double-click stapling of p53-derived peptides. ChemBioChem 15:2680–2683
61. Lau YH, de Andrade P, Skold N, McKenzie GJ, Venkitaraman AR, Verma C, Lanef DP, Spring DR (2014) Investigating peptide sequence variations for 'double-click' stapled p53 peptides. Org Biomol Chem 12:4074–4077
62. Lau YH, Wu YT, Rossmann M, Tan BX, de Andrade P, Tan YS, Verma C, McKenzie GJ, Venkitaraman AR, Hyvonen M, Spring DR (2015) Double strain-promoted macrocyclization for the rapid selection of cell-active stapled peptides. Angew Chem Int Edit 54:15410–15413
63. Hu K, Geng H, Zhang Q, Liu Q, Xie M, Sun C, Li W, Lin H, Jiang F, Wang T, Wu Y-D, Li Z (2016) An in-tether chiral center modulates the helicity, cell permeability, and target binding affinity of a peptide. Angew Chem Int Ed 55:8013–8017
64. Hou Z, Sun C, Geng H, Hu K, Xie M, Ma Y, Jiang F, Yin F, Li Z (2018) Facile chemoselective modification of thio-ethers generates chiral center-induced helical peptides. Bioconjug Chem 29:2904–2908
65. Hu K, Sun C, Li Z (2017) Reversible and versatile on-tether modification of chiral-center-induced helical peptides. Bioconjug Chem 28:2001–2007
66. Hu K, Sun C, Yu M, Li W, Lin H, Guo J, Jiang Y, Lei C, Li Z (2017) Dual in-tether chiral centers modulate peptide helicity. Bioconjug Chem 28:1537–1543
67. Vasco AV, Pérez CS, Morales FE, Garay HE, Vasilev D, Gavín JA, Wessjohann LA, Rivera DG (2015) Macrocyclization of peptide side chains by the ugi reaction: achieving peptide folding and exocyclic n-functionalization in one shot. J Org Chem 80:6697–6707
68. Assem N, Ferreira DJ, Wolan DW, Dawson PE (2015) Acetone-linked peptides: a convergent approach for peptide macrocyclization and labeling. Angew Chem Int Ed 54:8665–8668
69. Martín-Quirós A, Nevola L, Eckelt K, Madurga S, Gorostiza P, Giralt E (2015) Absence of a stable secondary structure is not a limitation for photoswitchable inhibitors of β-arrestin/β-adaptin 2 protein-protein interaction. Chem Biol 22:31–37
70. Zhang F, Timm KA, Arndt KM, Woolley GA (2010) Photocontrol of coiled-coil proteins in living cells. Angew Chem Int Ed 49:3943–3946
71. Bellotto S, Chen S, Rentero Rebollo I, Wegner HA, Heinis C (2014) Phage selection of photoswitchable peptide ligands. J Am Chem Soc 136:5880–5883
72. Miller SE, Kallenbach NR, Arora PS (2012) Reversible α-helix formation controlled by a hydrogen bond surrogate. Tetrahedron 68:4434–4437
73. Marqusee S, Baldwin RL (1987) Helix stabilization by Glu-...Lys + salt bridges in short peptides of de novo design. Proceedings of the National Academy of Sciences 84:8898–8902
74. Jackson DY, King DS, Chmielewski J, Singh S, Schultz PG (1991) General approach to the synthesis of short .alpha.-helical peptides. J Am Chem Soc 113:9391–9392
75. Toniolo C, Bonora GM, Bavoso A, Benedetti E, di Blasio B, Pavone V, Pedone C (1983) Preferred conformations of peptides containing α, α-disubstituted α-amino acids. Biopolym 22:205–215
76. Schafmeister CE, Po J, Verdine GL (2000) An all-hydrocarbon cross-linking system for enhancing the helicity and metabolic stability of peptides. J Am Chem Soc 122:5891–5892
77. Chen X-H, Xiang Z, Hu YS, Lacey VK, Cang H, Wang L (2014) Genetically encoding an electrophilic amino acid for protein stapling and covalent binding to native receptors. ACS Chem Biol 9:1956–1961

78. Leduc A-M, Trent JO, Wittliff JL, Bramlett KS, Briggs SL, Chirgadze NY, Wang Y, Burris TP, Spatola AF (2003) Helix-stabilized cyclic peptides as selective inhibitors of steroid receptor–coactivator interactions. Proc Natl Acad Sci 100:11273–11278
79. Galande AK, Bramlett KS, Trent JO, Burris TP, Wittliff JL, Spatola AF (2005) Potent inhibitors of LXXLL-based protein-protein interactions. ChemBioChem 6:1991–1998
80. Galande AK, Bramlett KS, Burris TP, Wittliff JL, Spatola AF (2004) Thioether side chain cyclization for helical peptide formation: inhibitors of estrogen receptor–coactivator interactions. J Pept Res 63:297–302
81. Sia SK, Carr PA, Cochran AG, Malashkevich VN, Kim PS (2002) Short constrained peptides that inhibit HIV-1 entry. Proc Natl Acad Sci 99:14664–14669
82. Harrison RS, Shepherd NE, Hoang HN, Ruiz-Gómez G, Hill TA, Driver RW, Desai VS, Young PR, Abbenante G, Fairlie DP (2010) Downsizing human, bacterial, and viral proteins to short water-stable alpha helices that maintain biological potency. Proc Natl Acad Sci 107:11686–11691
83. Taylor JW (2002) The synthesis and study of side-chain lactam-bridged peptides. Pept Sci 66:49–75
84. Baskin JM, Prescher JA, Laughlin ST, Agard NJ, Chang PV, Miller IA, Lo A, Codelli JA, Bertozzi CR (2007) Copper-free click chemistry for dynamic in vivo imaging. Proc Natl Acad Sci 104:16793–16797
85. Cantel S, Le Chevalier Isaad A, Scrima M, Levy JJ, DiMarchi RD, Rovero P, Halperin JA, D'Ursi AM, Papini AM, Chorev M (2008) Synthesis and conformational analysis of a cyclic peptide obtained via i to i + 4 intramolecular side-chain to side-chain azide – alkyne 1,3-dipolar cycloaddition. J Org Chem 73:5663–5674
86. Kawamoto SA, Coleska A, Ran X, Yi H, Yang C-Y, Wang S (2012) Design of triazole-stapled BCL9 α-helical peptides to target the β-catenin/B-Cell CLL/lymphoma 9 (BCL9) protein-protein interaction. J Med Chem 55:1137–1146
87. Torres O, Yüksel D, Bernardina M, Kumar K, Bong D (2008) Peptide tertiary structure nucleation by side-chain crosslinking with metal complexation and double "click" cycloaddition. ChemBioChem 9:1701–1705
88. Lau YH, de Andrade P, Quah ST, Rossmann M, Laraia L, Skold N, Sum TJ, Rowling PJE, Joseph TL, Verma C, Hyvonen M, Itzhaki LS, Venkitaraman AR, Brown CJ, Lane DP, Spring DR (2014) Functionalised staple linkages for modulating the cellular activity of stapled peptides. Chem Sci 5:1804–1809
89. Kim YW, Kutchukian PS, Verdine GL (2010) Introduction of all-hydrocarbon i, i + 3 staples into alpha-Helices via ring-closing olefin metathesis. Org Lett 12:3046–3049
90. Shim SY, Kim YW, Verdine GL (2013) A new i, i + 3 peptide stapling system for alpha-helix stabilization. Chem Biol Drug Des 82:635–642
91. Bird GH, Madani N, Perry AF, Princiotto AM, Supko JG, He XY, Gavathiotis E, Sodroski JG, Walensky LD (2010) Hydrocarbon double-stapling remedies the proteolytic instability of a lengthy peptide therapeutic. Proc Natl Acad Sci U S A 107:14093–14098
92. Hilinski GJ, Kim YW, Hong J, Kutchukian PS, Crenshaw CM, Berkovitch SS, Chang A, Ham S, Verdine GL (2014) Stitched alpha-helical peptides via bis ring-closing metathesis. J Am Chem Soc 136:12314–12322
93. Grossmann TN, Yeh JTH, Bowman BR, Chu Q, Moellering RE, Verdine GL (2012) Inhibition of oncogenic Wnt signaling through direct targeting of beta-catenin. Proc Natl Acad Sci U S A 109:17942–17947
94. Cui HK, Zhao B, Li YH, Guo Y, Hu H, Liu L, Chen YG (2013) Design of stapled alpha-helical peptides to specifically activate Wnt/beta-catenin signaling. Cell Res 23:581–584
95. Aurora R, Rosee GD (1998) Helix capping. Protein Sci 7:21–38
96. Koch O, Cole J (2011) An automated method for consistent helix assignment using turn information. Proteins: Struct, Funct, Bioinform 79:1416–1426
97. Penel S, Hughes E, Doig AJ (1999) Side-chain structures in the first turn of the α-helix1. J Mol Biol 287:127–143

98. Mahon AB, Arora PS (2012) End-capped α-helices as modulators of protein function. Drug Discov Today: Technol 9:e57–e62
99. Cabezas E, Satterthwait AC (1999) The hydrogen bond mimic approach: solid-phase synthesis of a peptide stabilized as an α-helix with a hydrazone link. J Am Chem Soc 121:3862–3875
100. Mahon AB, Arora PS (2012) Design, synthesis and protein-targeting properties of thioether-linked hydrogen bond surrogate helices. Chem Commun 48:1416–1418
101. Chapman RN, Dimartino G, Arora PS (2004) A highly stable short α-helix constrained by a main-chain hydrogen-bond surrogate. J Am Chem Soc 126:12252–12253
102. Kushal S, Lao BB, Henchey LK, Dubey R, Mesallati H, Traaseth NJ, Olenyuk BZ, Arora PS (2013) Protein domain mimetics as in vivo modulators of hypoxia-inducible factor signaling. Proc Natl Acad Sci 110:15602–15607
103. Henchey LK, Kushal S, Dubey R, Chapman RN, Olenyuk BZ, Arora PS (2010) Inhibition of hypoxia inducible factor 1—transcription coactivator interaction by a hydrogen bond surrogate α-helix. J Am Chem Soc 132:941–943
104. Henchey LK, Porter JR, Ghosh I, Arora PS (2010) High specificity in protein recognition by hydrogen-bond-surrogate α-helices: selective inhibition of the p53/mdm2 complex. ChemBioChem 11:2104–2107
105. Patgiri A, Yadav KK, Arora PS, Bar-Sagi D (2011) An orthosteric inhibitor of the Ras-Sos interaction. Nat Chem Biol 7:585–587
106. Hill TA, Shepherd NE, Diness F, Fairlie DP (2014) Constraining cyclic peptides to mimic protein structure motifs. Angew Chem Int Edit 53:13020–13041
107. Walensky LD, Kung AL, Escher I, Malia TJ, Barbuto S, Wright RD, Wagner G, Verdine GL, Korsmeyer SJ (2004) Activation of apoptosis in vivo by a hydrocarbon-stapled BH3 helix. Sci 305:1466–1470
108. Bernal F, Tyler AF, Korsmeyer SJ, Walensky LD, Verdine GL (2007) Reactivation of the p53 tumor suppressor pathway by a stapled p53 peptide. J Am Chem Soc 129:5298–5298
109. Moellering RE, Cornejo M, Davis TN, Del Bianco C, Aster JC, Blacklow SC, Kung AL, Gilliland DG, Verdine GL, Bradner JE (2009) Direct inhibition of the NOTCH transcription factor complex. Nat 462:152–187
110. Bird GH, Madani N, Perry AF, Princiotto AM, Supko JG, He X, Gavathiotis E, Sodroski JG, Walensky LD (2010) Hydrocarbon double-stapling remedies the proteolytic instability of a lengthy peptide therapeutic. Proc Natl Acad Sci 107:14093–14098
111. Phillips C, Roberts LR, Schade M, Bazin R, Bent A, Davies NL, Moore R, Pannifer AD, Pickford AR, Prior SH, Read CM, Scott A, Brown DG, Xu B, Irving SL (2011) Design and structure of stapled peptides binding to estrogen receptors. J Am Chem Soc 133:9696–9699
112. LaBelle JL, Katz SG, Bird GH, Gavathiotis E, Stewart ML, Lawrence C, Fisher JK, Godes M, Pitter K, Kung AL, Walensky LD (2012) A stapled BIM peptide overcomes apoptotic resistance in hematologic cancers. J Clin Invest 122:2018–2031
113. Brown CJ, Quah ST, Jong J, Goh AM, Chiam PC, Khoo KH, Choong ML, Lee MA, Yurlova L, Zolghadr K, Joseph TL, Verma CS, Lane DP (2013) Stapled peptides with improved potency and specificity that activate p53. ACS Chem Biol 8:506–512
114. Cui H-K, Zhao B, Li Y, Guo Y, Hu H, Liu L, Chen Y-G (2013) Design of stapled α-helical peptides to specifically activate Wnt/β-catenin signaling. Cell Res 23:581–584
115. Giordanetto F, Revell JD, Knerr L, Hostettler M, Paunovic A, Priest C, Janefeldt A, Gill A (2013) Stapled vasoactive intestinal peptide (VIP) derivatives improve VPAC(2) agonism and glucose-dependent insulin secretion. ACS Med Chem Lett 4:1163–1168
116. Leshchiner ES, Braun CR, Bird GH, Walensky LD (2013) Direct activation of full-length proapoptotic BAK. Proc Natl Acad Sci U S A 110:E986–E995
117. Long YQ, Huang SX, Zawahir Z, Xu ZL, Li HY, Sanchez TW, Zhi Y, De Houwer S, Christ F, Debyser Z, Neamati N (2013) Design of cell-permeable stapled peptides as HIV-1 integrase inhibitors. J Med Chem 56:5601–5612
118. Nomura W, Aikawa H, Ohashi N, Urano E, Metifiot M, Fujino M, Maddali K, Ozaki T, Nozue A, Narumi T, Hashimoto C, Tanaka T, Pommier Y, Yamamoto N, Komano JA, Murakami T, Tamamura H (2013) Cell-permeable stapled peptides based on HIV-1 integrase inhibitors derived from HIV-1 gene products. ACS Chem Biol 8:2235–2244

119. Okamoto T, Zobel K, Fedorova A, Quan C, Yang H, Fairbrother WJ, Huang DCS, Smith BJ, Deshayes K, Czabotar PE (2013) Stabilizing the pro-apoptotic BimBH3 Helix (BimSAHB) does Not necessarily enhance affinity or biological activity. ACS Chem Biol 8:297–302
120. Wei SJ, Joseph T, Chee S, Li L, Yurlova L, Zolghadr K, Brown C, Lane D, Verma C, Ghadessy F (2013) Inhibition of nutlin-resistant HDM2 mutants by stapled peptides, PLoS One 8
121. Frank AO, Vangamudi B, Feldkamp MD, Souza-Fagundes EM, Luzwick JW, Cortez D, Olejniczak ET, Waterson AG, Rossanese OW, Chazin WJ, Fesik SW (2014) Discovery of a potent stapled helix peptide that binds to the 70 N domain of replication protein A. J Med Chem 57:2455–2461
122. Spiegel J, Cromm PM, Itzen A, Goody RS, Grossmann TN, Waldmann H (2014) Direct targeting of rab-Gtpase-effector interactions**. Angew Chem Int Edit 53:2498–2503
123. Wang YX, Ho TG, Bertinetti D, Neddermann M, Franz E, Mo GCH, Schendowich LP, Sukhu A, Spelts RC, Zhang J, Herberg FW, Kennedy EJ (2014) Isoform-selective disruption of AKAP-localized PKA using hydrocarbon stapled peptides. ACS Chem Biol 9:635–642
124. Leshchiner ES, Parkhitko A, Bird GH, Luccarelli J, Bellairs JA, Escudero S, Opoku-Nsiah K, Godes M, Perrimon N, Walensky LD (2015) Direct inhibition of oncogenic KRAS by hydrocarbon-stapled SOS1 helices. Proc Natl Acad Sci U S A 112:1761–1766
125. Wang YX, Ho THG, Franz E, Hermann JS, Smith FD, Hehnly H, Esseltine JL, Hanold LE, Murph MM, Bertinetti D, Scott JD, Herberg FW, Kennedy EJ (2015) PKA-type I selective constrained peptide disruptors of AKAP complexes. ACS Chem Biol 10:1502–1510
126. Araghi RR, Ryan JA, Letai A, Keating AE (2016) Rapid optimization of Mcl-1 inhibitors using stapled peptide libraries including non-natural side chains. ACS Chem Biol 11:1238–1244
127. Teng Y, Bahassan A, Dong DY, Hanold LE, Ren XO, Kennedy EJ, Cowell JK (2016) Targeting the WASF3-CYFIP1 complex Using stapled peptides suppresses Cancer cell invasion. Cancer Res 76:965–973

# Chapter 2
# Synthesis of In-Tether Chiral Center Peptides and Their Biophysical Properties Study

## 2.1 Introduction

The majority of protein-protein interactions (PPIs) involve α-helices and are usually untreatable with small molecules due to their large interaction areas and shallow surfaces [1–3]. Therefore, the development of constrained peptides that can modulate PPIs by enhancing the helicity of short peptides is important [4–20]. Many strategies have been developed and successfully applied [21–25]. Early approaches to stabilize an α-helix involved both polar and physiological labile linkages, such as disulfide bonds [7] and lactam bridges [5, 26]. A significant advance in this field was made by applying ring-closing metathesis to construct constrained peptides [27]. Verdine, Walensky, and others developed a hydrocarbon staple that significantly increased the peptide's α-helical content, proteolytic resistance, and enhanced biological activities [10, 28–30]. This strategy has been used to perturb protein-protein interactions such as intracellular Bcl-2 [28], MDM2 [31, 32], and extracellular EGFR [33]. Arora et al. developed a hydrogen bond surrogate system that was also broadly utilized [34–37]. Due to the therapeutic potential of constrained peptides, alternative peptide 'stapling' methods were developed, including hydrozone by Satterthwait [8], azobenzene by Woolley [9], thioether by Spatola [12], perfluoroaryl formation by Pentelute [18], oxime formation by Brown [38], and other cysteine-based $S_NAr$ by Qin and others [39, 40]. These strategies are summarized in Scheme 2.1. These special linkers increase peptide backbone rigidity that leads to a more rigid helical conformation.

The solved crystal structures of estrogen receptor-α (ER-α) and E3 ubiquitin ligase MDM2 with their peptide ligands show that the hydrocarbon tethers contribute to target binding. This is due to that the tether interacts with the hydrophobic region surrounding the binding pocket [31, 41]. Incorporating a modification site on the tether could lead to versatile applications. This concept was first demonstrated by Dawson et al. [42]. They built a carboxyl group into the linker for further oxime ligation to be used for fluorescent molecule labeling. Smith et al. utilized the inverse electron demand Diels–Alder reaction of the S, S-tetrazine for peptide stapling and



K. Hu, *Development of In-Tether Carbon Chiral Center-Induced Helical Peptide*, Springer Theses,
https://doi.org/10.1007/978-981-33-6613-8_2

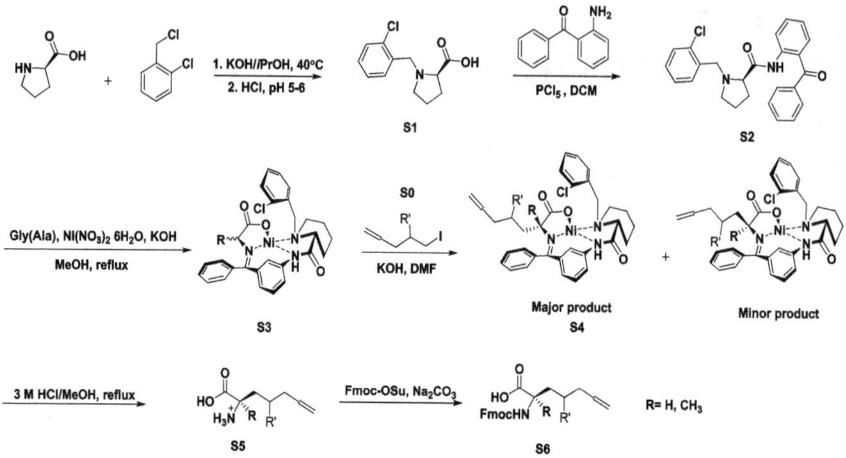

**Scheme 2.1** Flowchart for unnatural amino acids synthesis (*Note* The R' group in $S_0$ is changeable)

labeling [43]. However, the influences of above-mentioned methods on the peptides' biophysical properties have not been explored in detail.

Cell permeability is one of the major limiting factors for peptide therapeutics that targeting intracellular targets. Cell permeability can be modulated by many factors such as peptide conformation [44–46]. However, while the parameters that determine a peptides' permeability would be valuable for developing cell permeable peptides, these parameters remain poorly understood. Recently Verdine et al. have systematically analyzed more than 200 peptides and provided valuable information for understanding the permeability of peptides [47]. They observed that the permeability of stapled peptides is influenced by the staple type and the formal charge or the peptide sequence rather than other physicochemical parameters. However, their study remains some limitations. It should be noted that scrambling the positions of a few amino acids in a peptide would dramatically change its permeability and other biophysical properties. Thus, finding the proper controls to evaluate a peptide's biophysical properties is still a formidable task.

We wondered whether a chiral center in the tether of a stapled peptide, as shown in Fig. 2.1, could influence the secondary structure and physical property of the peptide. In this chapter, we synthesized a series of stapled peptides containing a carbon atom chiral center within the tether. We have found that a precisely positioned chiral center significantly improves the α-helical contents, protease resistance and cell permeability. We also found that the chiral center can modulate target binding affinity. Thus, these peptides provide an ideal platform to investigate the change in a peptide's biophysical properties caused by its conformational change.

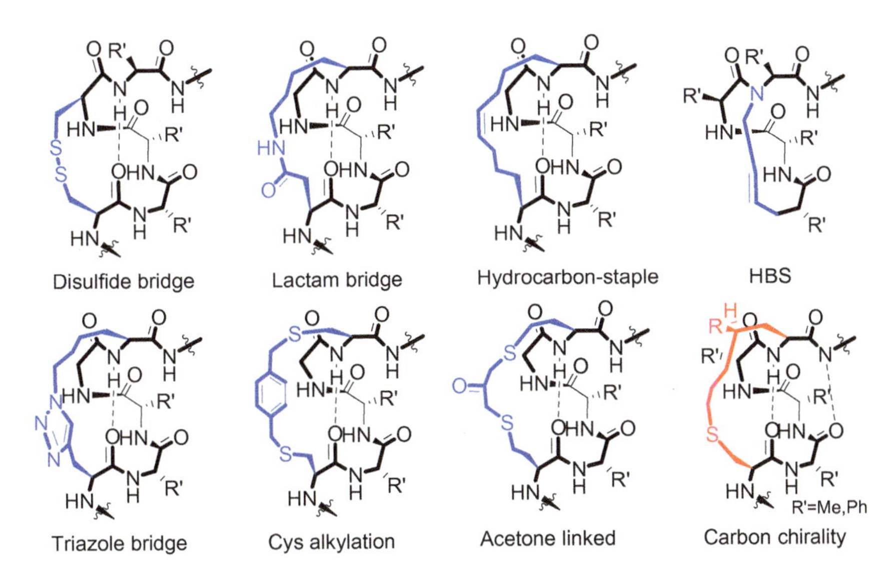


**Fig. 2.1** The α-helical peptide constrained strategies developed by others and by our group

## 2.2 Results

### 2.2.1 *Unnatural Amino Acids Synthesis*

All the unnatural amino acids were synthesized based on the previous literatures [48–52], as shown in Scheme 2.1. The detailed synthesized procedures for $S_5$(2-Me) and $S_5$(2-Ph) were described below. For other nonnatural amino acids, we synthesized different bromide substituted side chains and use the same procedure to synthesize the target compounds. The chemical structures of all nonnatural amino acids are shown in Scheme 2.2.

**Compound S1**: Potassium hydroxide (38.4 g, 0.7 mol) was dissolved in anhydrous methanol (125 ml) and heated to 60°C, then D-proline (23 g, 0.2 mol) was added into the mixture. After complete dissolution, 2-chlorobenzyl chloride (32 ml, 0.26 mol) was added dropwise. After 24 h, $CH_2Cl_2$ (100 ml) was added and the reaction mixture stood for 4 h. Then the mixture was filtered out and the residue was washed by $CH_2Cl_2$ twice. The filtrate was gathered, concentrated, and crystallized in acetone to obtain compound (41 g, yield: 83%).

**Compound S2**: Compound 1 (24.8 g, 0.1 mol) was added into $CH_2Cl_2$ (200 ml) and the mixture was cooled to 0°C. Phosphorus pentachloride (30.2 g, 0.15 mol) was added dropwise and stirred for 1 h, followed by the addition of 2-aminobenzophenone (20.0 g, 0.1 mol). The reaction was stirred at r.t. for 4 h. $CH_2Cl_2$ was removed under vacuum and acetone was added for crystallization to obtain compound 2 (25.2 g, yield: 61%).

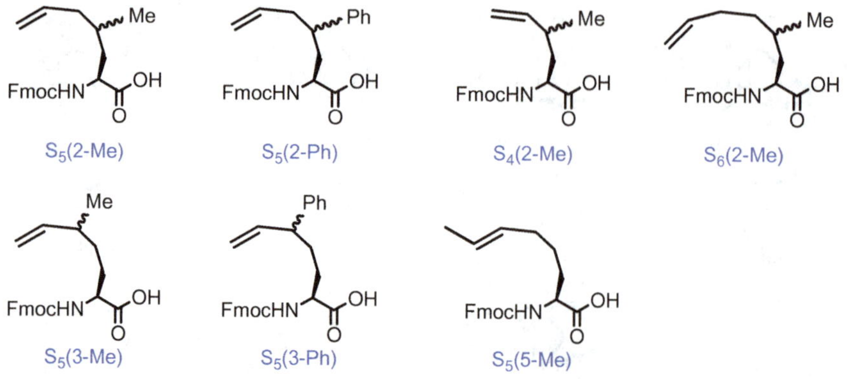

$S_5$(2-Me) $S_5$(2-Ph) $S_4$(2-Me) $S_6$(2-Me)

$S_5$(3-Me) $S_5$(3-Ph) $S_5$(5-Me)

**Scheme 2.2** Chemical structures of nonnatural amino acids

**Compound S3**: Compound 2 (25.0 g, 0.065 mol), nickel (II) nitrate hexahydrate (31.6 g, 0.11 mol) and glycine (20.5 g, 0.27 mol) were dissolved in anhydrous methanol (300 ml) and heated to 50°C. The potassium hydroxide (25.0 g, 0.47 mol) in methanol (150 ml) solution was added dropwise. After 10 h, acetic acid was added. Methanol was removed and followed by pouring the residue liquid into ice water (800 ml), and stirred at r.t. overnight to promote precipitation. The mixture was filtered out under vacuum and residue was gathered to obtain red solid compound 3 (22 g, yield: 75%).

**Compound S4**: Under $N_2$ atmosphere, compound 3 (20.0 g, 0.04 mol) was dissolved in DMF (200 ml), followed by the addition of powdered potassium hydroxide (21.1 g, 0.4 mol) and the reaction mixture was stirred at r.t. for 1 h. Under the condition of ice bath, 5-bromo-1-pentene (6 ml, 0.042 mol, J&K Co. Ltd) was added dropwise. Then the reaction was gradually warmed to r.t. and stirred for 4 h before the addition of 5% v/v acetic acid in water. The reaction continued to be stirred for 6 h to promote the precipitation and filtered out. The residue was gathered and washed by water for three times to obtain compound 4 (23.2 g, yield: 87%).

**Compound S5**: Compound 4 (23.2 g, 0.035 mol) was dissolved in methanol/$CH_2Cl_2$ (v/v = 50 ml/100 ml), and 3 M hydrochloric acid (100 ml) was added into the mixture. The reaction was heated to 80°C and stirred overnight until yellow/green color change was observed. Then the solvent was removed in vacuo and chloroform was used for extraction for three times to recover the ligand. The amino acid aqueous fraction was used for the next step without further purification.

**Compound S6**: Sodium bicarbonate (16.8 g, 0.2 mol) and EDTA-Na (18.6 g, 0.05 mol) were added into the aqueous fraction to remove residual nickel. After stirring for 20 min, sodium bicarbonate was added again to make pH value of the mixture stay at 7–8. Then the mixture was cooled to 0°C with ice bath. 9-fluorenylmethyl succinimidyl carbonate (11.7 g, 0.035 mol) was dissolved in acetonitrile (25 ml) and added dropwise into the aqueous solution. The reaction was gradually warmed to r.t. and stirred for 12 h. Acetonitrile was removed in vacuo and citric acid was added to make pH value of the mixture stay at 2–3. The reaction was extracted with ethyl acetate for three times. The organic phase was dried with anhydrous magnesium sulfate. The final product $S_6$ was obtained after the purification of flash chromatography (Hexane: EA = 5:1) (5.2 g, yield: 41%)

### 2.2.2 *Solid Phase Peptide Synthesis and Thiol-Ene Photoreactions*

Peptides were synthesized on MBHA resin (loading capacity: 0.37 mmol/g) by standard Fmoc-based SPPS (Scheme 2.3). At first, the resin was swelled in NMP for 30 min. Then 50% (vol/vol) morpholine in NMP was used to deprotect the Fmoc group on amine group for 30 min × 2. Next the resin was washed with DCM and NMP alternatively for three time (3×1 min). In coupling process, for natural amino acids, the Fmoc-protected amino acids (5.0 equiv), HCTU (4.9 equiv), DIPEA (10.0 equiv) were dissolved in NMP and mixed with resin for 2 h, followed by washing with DCM and NMP for three times (3×1 min). For unnatural amino acids, Fmoc-protected acids (2.5 equiv), HCTU (2.4 equiv), and DIEA (5.0 equiv) were dissolved in NMP and mixed with resin for 4 h, followed by washing with DCM and NMP for several times. After we completed the synthesis of designed peptide, the intramolecular thiol-ene reaction was performed. The resins were drained and transferred to a suitable flask, mixed with 1.2 eq MAP/MNP (1:1) catalyst in DMF and reacted at ultraviolet light (365 nm) for 2 h. Final the resins were treated with a mixture of TFA:$H_2O$:TIS (95:2.5:2.5 by volume) for 2 h and dried by blowing nitrogen. Then the peptides were precipitated with Hexane:$Et_2O$ (1:1 in volume) at 4°C, isolated by centrifugation and dissolved in 40% (vol/vol) acetonitrile/water, purified by HPLC with UV absorbance at 220 nm or 280 nm and later analyzed by LC-MS.

### 2.2.3 *Circular Dichroism Study of Secondary Structures*

To eliminate possible amino acid residue perturbations, a single turn pentapeptide system was employed as a model system based on previous literature [13]. In order to access the linear peptide and both cyclic peptide diastereomers in one synthesis, the unnatural amino acid epimers were used. The structure-activity relationship of the tether ring sizes and chiral center positions are summarized in Scheme 2.4 and Fig. 2.2 and the optimal tether is shown in Fig. 2.3. These results suggested that a seven-atom tether with a C-terminal $\gamma$-position chiral center induces the highest helical content of peptides. Cyclic peptides **1a** and **1b** were synthesized from the linear peptide **1**, cyclo-Ac-CAAA$S_5$(2-Me)-$NH_2$ ($S_5$: (S)-pentenylglycine, 2-Me: the methyl group located at the $\beta$ position to the $\alpha$ carbon of amino acid), via a thiol-ene reaction as shown in Fig. 2.3a. Peptides **1a** and **1b** were readily separable by

**Scheme 2.3** The representative synthesis procedures for pentapeptides (MAP: 4-methoxyacetophenone, MMP: 2-hydroxy-4'-(2-hydroxyethoxy)-2-methylpropiophenone, DIEA: N, N-diisopropylethylamine), TIS: Triisopropylsilane

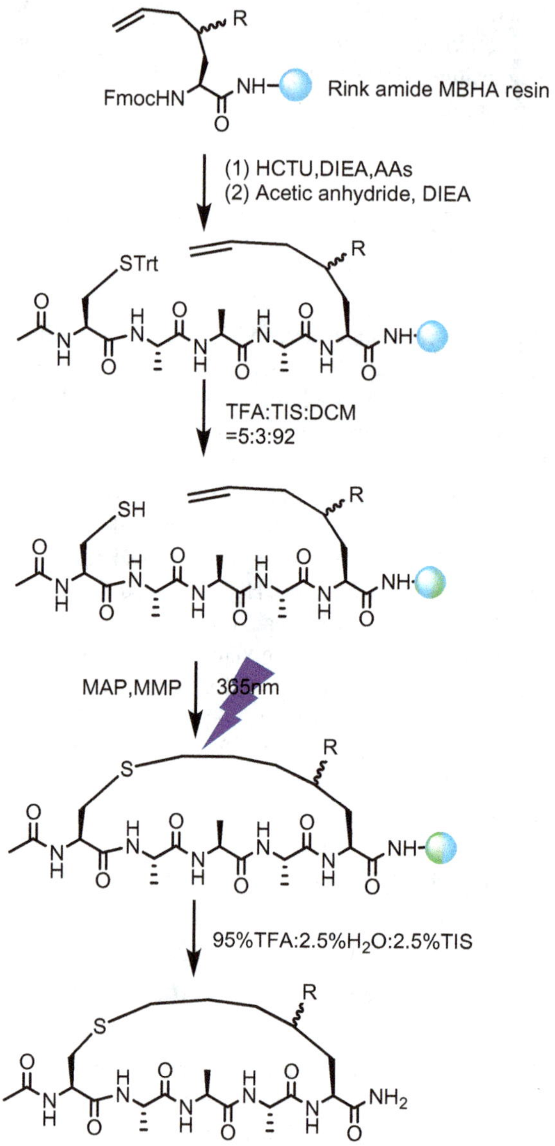

FmocHN NH- Rink amide MBHA resin R O
(1) HCTU,DIEA,AAs
(2) Acetic anhydride, DIEA
STrt R NH
TFA:TIS:DCM =5:3:92
SH R NH
MAP,MMP 365nm
S R NH
95%TFA:2.5%$H_2O$:2.5%TIS
S R $NH_2$

reverse-phase HPLC, suggesting significant conformational differences in solution. All peptides categorized in group **b** have longer retention times when compared to their diastereomers in group **a**. Circular dichroism (CD) spectroscopy measurements clearly show that peptide **1a** is a random coil while **1b** is helical in PBS buffer (PH = 7.0, Fig. 2B).

A series of single turn peptides with varying sequences were tested and all peptide diastereomers (peptides **2a/2b** to peptides **10a/10b**) were easily separable by HPLC.

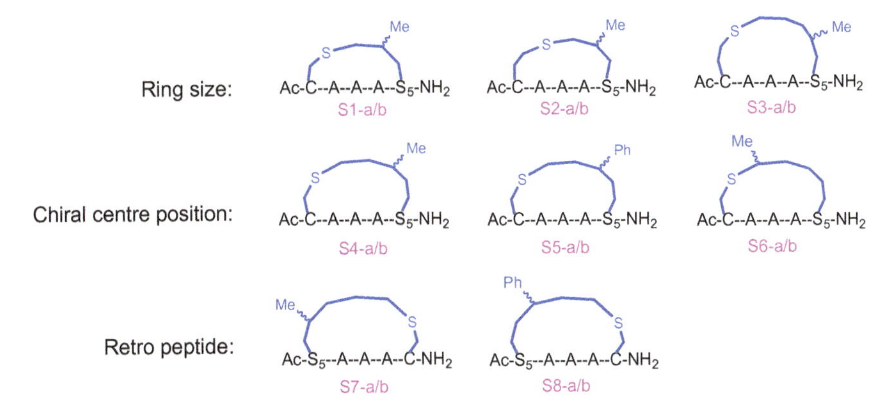


**Scheme 2.4** Illustration of pentapeptides with variable ring sizes and chiral center positions

In all cases, including peptide **8a/8b**, which contains a glycine residue, the **b** diastereomers show enhanced helicity while the **a** diastereomers were mainly random coils. These results are summarized in Fig. 2.3 and Table 2.1.

We then examined stability of the peptides. Peptide **1b** remains helical at high temperature and at high concentrations of guanidinium hydrochloride, which suggests that the helix is stabilized by the in-tether chiral center (Fig. 2.4). Notably, peptide **2b** shows increased helicity over peptide **1b**, indicating that there may be a benefit to larger substitution groups.

### *2.2.4 NMR Study of Peptides' Secondary Structures*

To further understand the effect of in-tether chiral center constraint on peptide conformational preference in aqueous solution, a detailed 1D and 2D $^1$H-NMR study of **1b**, **2b**, and **10b** was performed in 10% $D_2O$ in $H_2O$ at 25°C. 2D-TOCSY spectra were used to identify resonances for each amino acid (Table 2.2), and NOESY spectra were used to identify sequential connectivity and intraresidue NH-NH and NH-CH cross-peaks. As expected, there were a number of spectra features that are well characteristics of a well-defined structure in the cyclic pentapeptides and specifically characteristics of alpha helicity except the C termini residue $S_5$(2-Me/2-Ph). Firstly, there were conspicuously low coupling constants ($^3J_{NH\text{-}Cha} < 6$ Hz) for amide resonances except $S_5$(2-Me/2-Ph) (Fig. 2.5 and Table 2.3), as normally observed in α-helical peptides [53]; secondly, the observation in NOESY spectra of nonsequential medium range $d_N(i, i+3), d\ (i, i+3),$ *and* $d_N(i, i+4)$ NOEs suggest helical structure (Fig. 2.5). Furthermore, the temperature coefficients of the backbone amides NH chemical shifts of **1b** was determined, with temperature coefficients (Δδ/T) being < 4 ppb/K for C5 and A2 (Fig. 2.6), consistent with their involvement in hydrogen bonds that characterize an -helix or $3^{10}$ helix. In summary, the NOE spectrums and

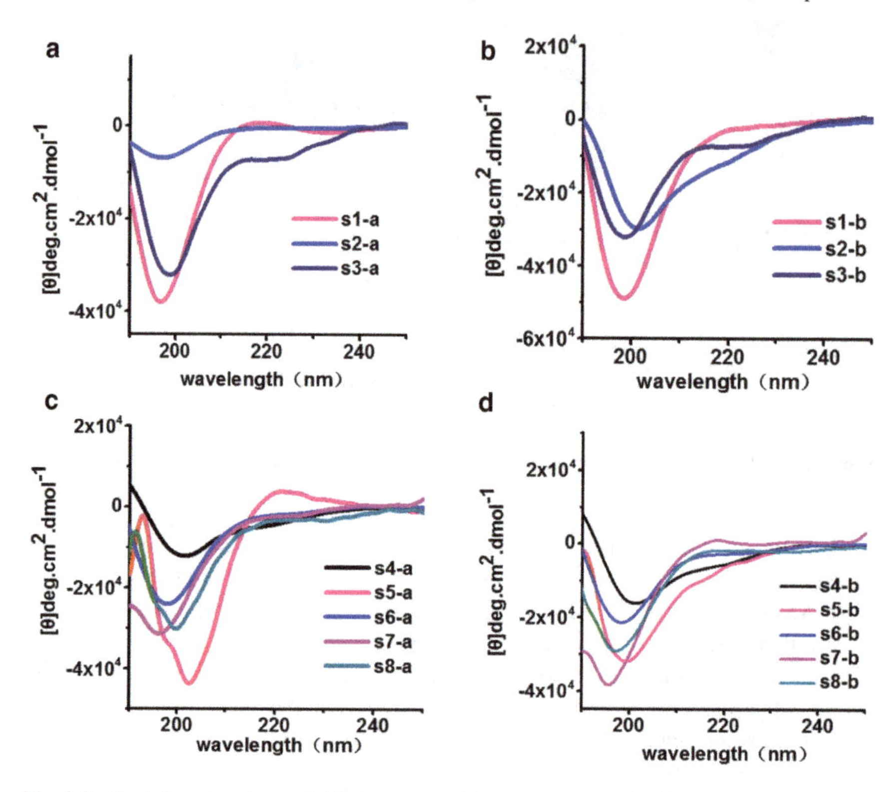


**Fig. 2.2** The tether ring size and chiral center position were systematically studied. Peptides with side chain atom number 5–8 were synthesized, their CD spectra were shown in (A, B). The a(s) were designated as the S diastereomers and the b were designated as the R diastereomers. The optimal ring size for most helicity is 7 atoms of the crosslinker (C, D). CD spectra of peptides with different chiral center positions. C for the S diastereomers and D for the R diastereomers. All measurements were performed in PBS (PH = 7.0) in 20°C. The chiral center position, 3-Me (black), 3-Ph (red), 5-Me (blue), retro peptides: 2-Me (pink), 3-Ph (green). For peptides with 1' or 4' position chiral center, we noticed that these two nonnatural amino acids were very difficult to synthesis or highly unstable, so they were out of our optimized table

CD spectra of peptides **1b**, **2b** and **10b** suggest they adopt a mixture of $3^{10}$ helix and -helix in solution. This conclusion is further validated by single crystal structure of **10b** and computational studies.

### 2.2.5 *X-Ray Diffraction Analysis of the Crystal Structure*

As shown in Fig. 2.7 and Table 2.4, X-ray diffraction analysis of peptide **10b** cyclo-Ac-CAAIS$_5$(2-Me)-NH$_2$ unambiguously confirms that the absolute configuration of the in-tether chiral center is *R*. The intramolecular hydrogen bonding pattern (indicated by the dashes bonds, Table 2.5) agrees well with the previously proposed

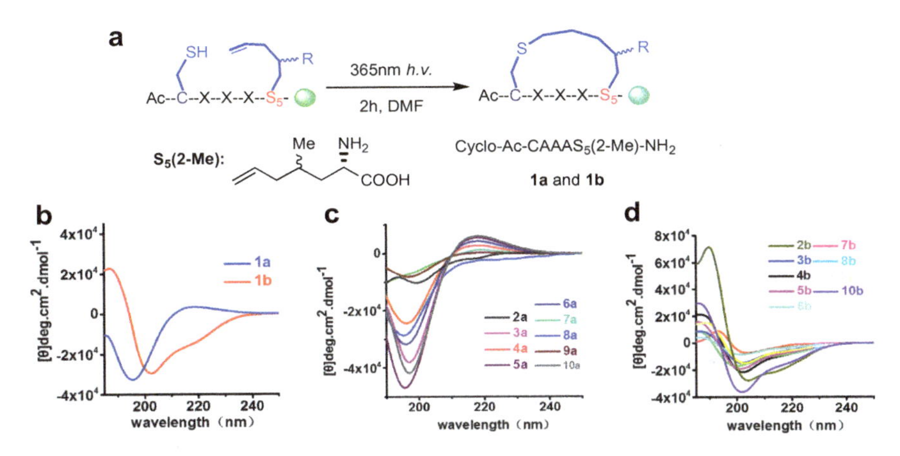


**Fig. 2.3** Helicity enhancements with an in-tether chiral center. (A) Schematic presentation of constrained peptide preparation. CD spectra of cyclic pentapeptides 1a/1b (B), 2a-10a (C), and 2b-10b (D) in PBS (PH = 7.0) at 20°C

**Table 2.1** Molar ellipticities and percentage of helicity of peptides 1b-10b in PBS (PH = 7.0) at 20°C. (*): The α-helical content of each peptide was calculated based on the method reported previously. The final helical content presented as relative to peptide **2b**, as fixed the peptide **2b** as 100% helicity

| Entry | Peptide | [θ]222 | [θ]205 | [θ]190 | Helicity |
|---|---|---|---|---|---|
| 1b | cyclo-CAAAS5(2-Me) | −11792 | −27010 | 17974 | 0.87 |
| 2b | cyclo-CAAAS5(2-Ph) | −13780 | −27001 | 70588 | 1 |
| 3b | cyclo-CAIAS5(2-Me) | −4415 | −13138 | 6425 | 0.37 |
| 4b | cyclo-CAEAS5(2-Me) | −6784 | −19340 | 17519 | 0.53 |
| 5b | cyclo-CASAS5(2-Me) | −5761 | −16485 | 9514 | 0.46 |
| 6b | cyclo-CAQAS5(2-Me) | −2682 | −7651 | 2403 | 0.26 |
| 7b | cyclo-CAFAS5(2-Me) | −977 | −7170 | 5469 | 0.14 |
| 8b | cyclo-CAGAS5(2-Me) | −6931 | −14272 | 3080 | 0.54 |
| 9b | cyclo-CEAKS5(2-Me) | −4751 | −13750 | 14232 | 0.4 |
| 10b | cyclo-CAAIS5(2-Me) | −12139 | −32516 | 23162 | 0.9 |

helix model as shown in Fig. 2.1. The backbone dihedral angle set is summarized in Table 2.6, and all dihedral angle values are close to that of a standard α-helix except for the C terminal residue. In addition, the methyl group at the chiral center protrudes from the peptide backbone. Therefore, a chiral center in the tether provides a modifiable site which can lead to more effective peptide ligands or be utilized to improve the peptide's drug-like property.

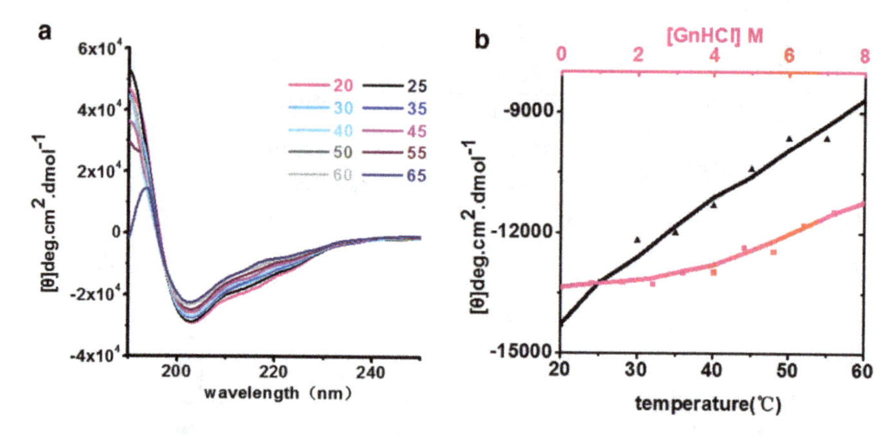


**Fig. 2.4** Stability tests of peptide **2b**. (**a**) Dependence on temperature (20–65°C) of full CD spectra of **2b in PBS (PH = 7.0)**. (**b**) Mean residue ellipticity at 222 nm of **2b** at different temperatures (black line). Variation in molar ellipticity of **2b** at 222 nm with increasing [guanidine HCl] at 25°C (red line)

**Table 2.2** $^1$H-NMR chemical shifts (δ, ppm) for peptide **1b**, **2b**, and **10b** in $H_2O$ with 10% $D_2O$ at 298 K

| No | Residue | NH | Hα | Hβ | H (sidechain) |
|---|---|---|---|---|---|
| *Peptide **1b*** | | | | | |
| 1 | $S_5$(2-Me) | 8.04 | 4.29 | – | 1.62, 1.46, 1.27, 0.74 |
| 2 | A | 7.96 | 4.12 | 1.30 | – |
| 3 | A | 7.73 | 4.18 | 1.27 | – |
| 4 | A | 8.49 | 4.11 | 1.29 | – |
| 5 | C | 8.33 | 4.30 | 2.85/2.78 | – |
| *Peptide **2b*** | | | | | |
| 1 | $S_5$(2-Ph) | 8.19 | 4.13 | – | 3.66, 2.89, 2.12, 1.90, 1.68, 1.29 |
| 2 | A | 7.99 | 4.13 | 1.47 | – |
| 3 | A | 7.83 | 4.20 | 1.37 | – |
| 4 | A | 8.48 | 4.15 | 1.35 | – |
| 5 | C | 8.33 | 4.42 | 3.02/2.79 | – |
| *Peptide **10b*** | | | | | |
| 1 | $S_5$(2-Me) | 8.09 | 4.3 | – | 1.61, 1.41, 1.27, 1.17, 0.74 |
| 2 | I | 7.78 | 3.94 | 1.77 | 1.44, 1.11, 0.80 |
| 3 | A | 7.66 | 4.21 | 1.28 | – |
| 4 | A | 8.56 | 4.14 | 1.3 | – |
| 5 | C | 8.3 | 4.26 | 2.87/2.79 | – |

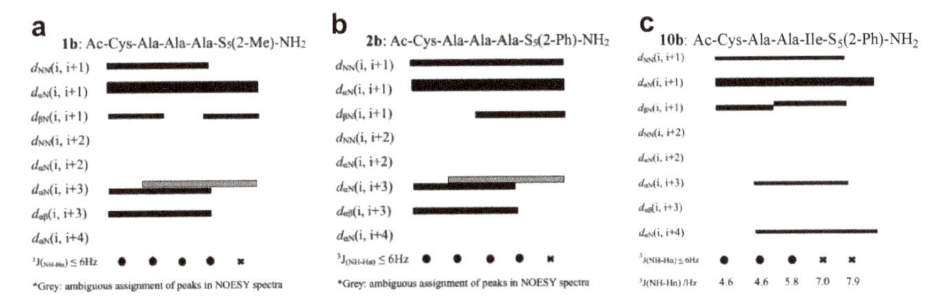


**Fig. 2.5** NOE summary diagram for peptide **1b** (A), **2b** (B), and 10b (C) in 90%$H_2O$:10%$D_2O$ at 298 K. Thickness of bars reflects intensity of NOEs. Bar thickness reveals the intensity of the NOE signals

**Table 2.3** $^3J_{NH\text{-}H}$ (HZ) for peptide **1b**, **2b**, and **10b** in $H_2O$ with 10% $D_2O$ at 298 K

| $^3J_{NH\text{-}H}$ (Hz) | | | | | |
|---|---|---|---|---|---|
| **1b** | $S_5$(2-Me)1 | A2 | A3 | A4 | C5 |
| | 7.9 | 4.7 | 5.6 | 4.3 | 4.6 |
| **2b** | $S_5$(2-Ph)1 | A2 | A3 | A4 | C5 |
| | 8.4 | 5.4 | 5.8 | 4.4 | 4.5 |
| **10b** | $S_5$(2-Me)1 | 12 | A3 | A4 | C5 |
| | 7.9 | 7 | 5.8 | 4.6 | 4.6 |

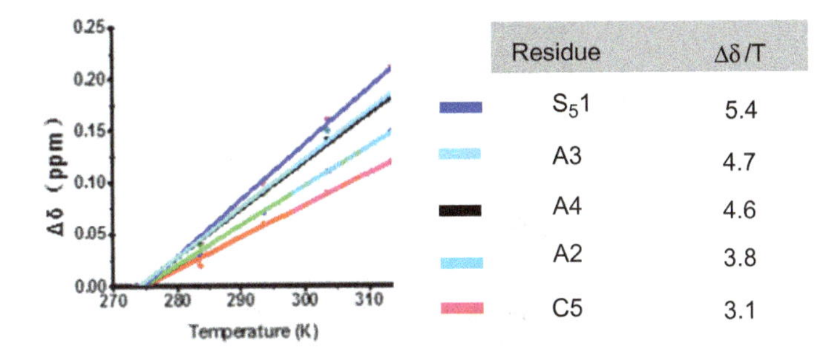


**Fig. 2.6** Temperature dependence (Δδ/T, ppb/K) for amide NH chemical shift of peptide **1b**

### 2.2.6 *Molecular Dynamic Simulations*

In order to better understand the conformational features of different peptide diastereomers, we performed replica exchange molecular dynamics (REMD) simulations with explicit water using the recently developed forcefield RSFF2. For peptide **10b**, the most distributed structure derived from the simulation is almost identical to the solved structure (Main chain + Cβ rmsd: 0.3 Å) as shown in Fig. 2.8a. The Ramachandran plots (, distribution) of the two diastereomers in the simulations exhibit different

**Fig. 2.7** Thermal ellipsoid and backbone H-bonds crystal structures of pentapeptide **10b** cyclo-Ac-CAAIS$_5$(2-Me)-NH$_2$

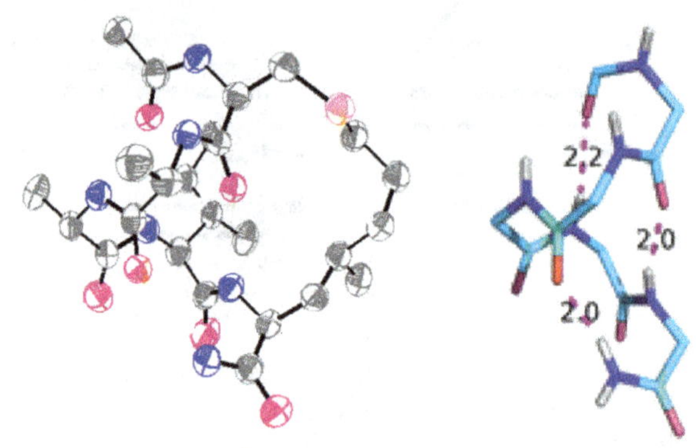


**Table 2.4** Statistics of data collection and structure refinement

| Crystal name | Cyclo-Ac-CAAIS$_5$(2-Me)-NH$_2$ |
|---|---|
| **Data collection** | Cu Kα |
| Chemical formula<br>Molecular Weight<br>Temperature(K)<br>Wavelength(Å)<br>Space group | C25 H44 N6 O6 S<br>556.72<br>100<br>1.5418<br>P212121 |
| *Cell dimensions* | |
| $a, b, c$ (Å) | 8.94, 19.40, 20.42 |
| $\alpha, \beta, \gamma$ (°) | 90.0, 90.0, 90.0 |
| Resolution (Å) | 20.42-0.82(0.86-0.82) |
| $R_{merge}$ | 0.15(0.41) |
| $I/\sigma I$ | 13.6(3.0) |
| Completeness (%) | 99.1(94.0) |
| Redundancy<br>Mosaicity | 9.72(3.44)<br>1.32 |
| *Refinement* | |
| Resolution (Å)<br>Refinement method | 0.82<br>Full-matrix least-squares on $F^2$ |
| No. Reflections Measured<br>Unique reflections<br>Corrections<br>Friedel pairs<br>Structure solution<br>Flack parameter | 36,111<br>6500<br>Lorentz-polarization<br>2805<br>Direct Methods (ShelxD)<br>0.19 |
| R [F2 > 2σ(F2)], wR(F2)* | 0.23, 0.58 |
| Goodness-of-fit on $F^2$ | 1.867 |
| H-atom treatment | H atom Parameters constrained |

Values in parentheses refer to data in the outlier resolution shell
*High values beyond of normal range due to the difficulty of crystallization thus bad crystals have to be used for data collection

**Table 2.5** Hydrogen bonds parameters of peptide crystal. The distance of H…O atoms at the i, i + 4 positions is less than 2.8 Å, which means forming strong intramolecular hydrogen bond

| Parameters of the hydrogen bonds (Å,°) | | | | | |
|---|---|---|---|---|---|
| Donor | Acceptor | N…O | H…O | N-H…O | C-O…H |
| N-NH2 | O4 | 2.84 | 2.03 | 152 | 143 |
| N1 | O5 | 2.88 | 2.02 | 163 | 156 |
| N2 | O-Ac | 3.05 | 2.22 | 158 | 156 |

**Table 2.6** The dihedral angles of solved crystal. 4 of 5 pairs of parameters except the $S_5$ closed to standard alpha helix means **10b** formed helix in $H_2O$

| Residue | $\varphi$ | $\psi$ | $\omega$ |
|---|---|---|---|
| $S_5$1 | −92 | −22 | −166 |
| I2 | −68 | −53 | −180 |
| A4 | −62 | −38 | −177 |
| A4 | −63 | −42 | −179 |
| C5 | −57 | −42 | −130 |

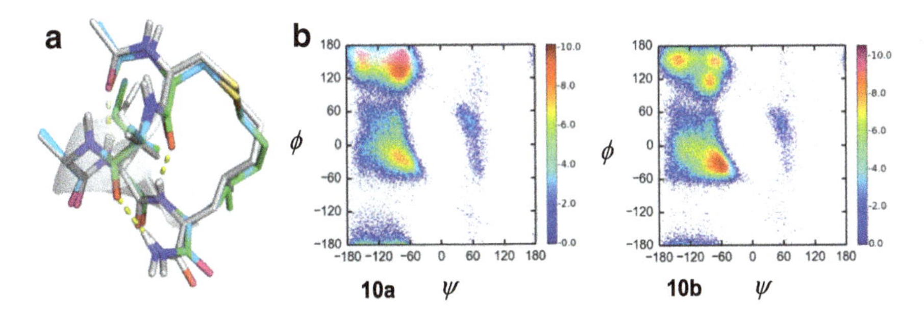


**Fig. 2.8 Conformation analysis of peptide 10a/10b cyclo-Ac-CAA IS$_5$(2-Me)-NH$_2$**. (C) Calculated structure of peptide **10b** superimposed with solved structure. Each simulation ran over 200 ns for sufficient sampling. Snapshots using to analysis were taken from replica at room temperature (300 K) and conformational clustering conduct using a backbone dihedral-based method. (D) Ramachandran plots of **10a/b** from REMD simulation. Left (**10a**), right (**10b**)

conformational preferences (Fig. 2.8b). For the S-diastereomer **10a**, the dominant calculated structures are shown in Figs. 2.9 and 2.10, and demonstrate no significant secondary structures, which is in excellent agreement with CD results. Further simulation of a peptide without the in-tether R-substitution group (Ac-cyclo-CAAAS$_5$(2-H)-NH$_2$) indicates that the polyproline-II (PII) conformation is intrinsically favored by the residues, and the representative structure of most populated cluster is not helical (Fig. 2.9a). In this non-helical structure, a R = $CH_3$/Ph substitution with (S)-chirality can be added without any steric interference (Fig. 2.9b). However, the non-helical structure will be significantly destabilized when a R = $CH_3$ substitution group is placed in (R)-chirality, and it will be very comfortable when the peptide

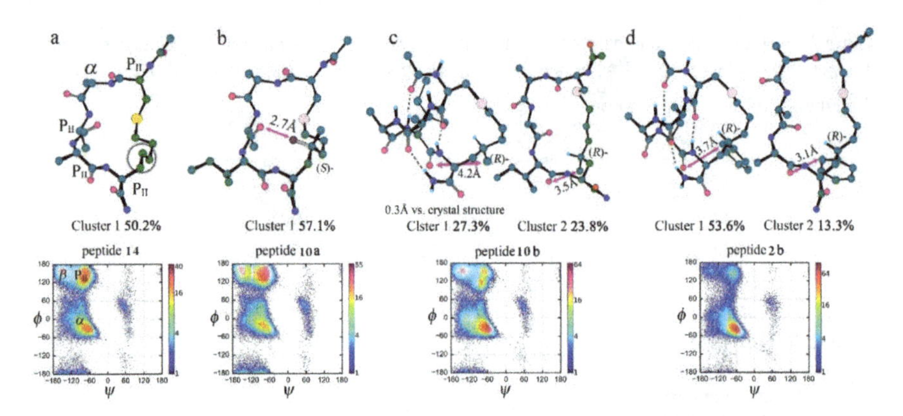


**Fig. 2.9 Representative structures and Ramachandran ($\phi$, $\psi$) plots from REMD simulations of peptides 14 (a), 10a (b), 10 (c), 2b (d).** The Cluster 1 and Cluster 2 indicate the most populated and second most populated structure clusters, respectively. All hydrogen atoms are omitted for clarity, except for the polar H atoms in -helical structures and the tertiary H atom on the chiral center in each tether. Each $\phi$, $\psi$ plot was drawn using the conformations of all five residues together, and each color bar indicates relative probability in logarithmic scale. (a) The backbone conformation of each residue is labeled, and the potential substitution site in the tether is stressed by a circle. (b) The peptide has (*S*)-tether but a hypothetic (*R*)-CH3 substitution was also placed in brown color to illustrate the steric clash

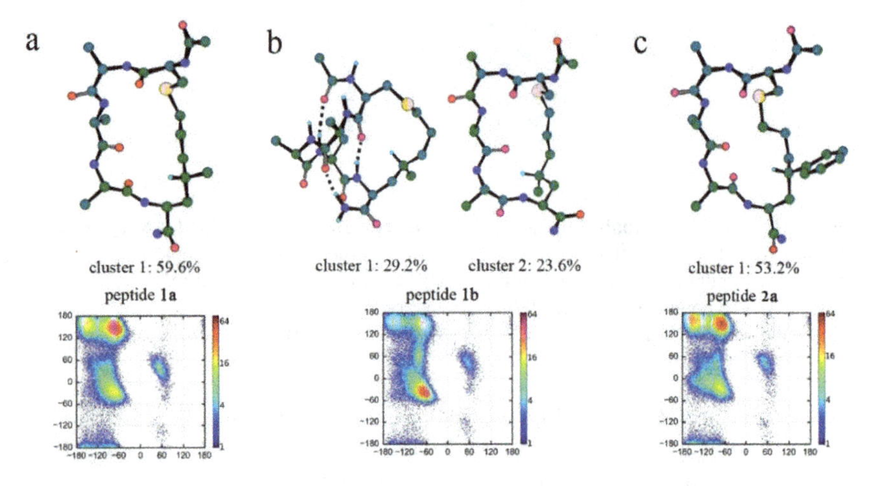


**Fig. 2.10** The representative structures of **1a**, **1b**, and **2a**, and their Ramachandran ($\phi$, $\psi$) plots from simulation

backbone adopts α-helical conformation (Fig. 2.9c). Larger R = Ph group in (R)-chirality will lead to higher destabilization of non-helical structures, and stronger preference for α-helical conformation (Fig. 2.9d).

### 2.2.7 On the Mechanism of Chirality-Induced Helicity

To better understand the effects of the in-tether chiral center on the -helix content, we carried out simulations on peptides **14** Ac-cyclo-CAAAS$_5$-NH$_2$ (R = H), **10a/10b** (R = CH3), and **2a/2b** (R = Ph). Peptides **1a/1b** were also simulated (Fig. 2.10a, b), and results very similar to those of **10a/10b** were obtained (Fig. 2.9). Here we use **10a/10b** to discuss because only **10b** has the crystal structure. By using replica-exchange molecular dynamics (REMD) method, converged conformational sampling was achieved for each peptide. The simulated results are in good agreement with experimental observations. For peptide **14**, the representative structure is non-helical, with four out of five residues in PII conformation (Fig. 2.9a). Indeed, from its Ramachandran ($\phi$, $\psi$) plot, PII is more favored than the and β conformations. This can be understood by the recent findings that PII conformation is intrinsically preferred for short unfolded peptides in aqueous solution. The representative structure of peptide **10a** with (S)-chirality in the tether is very similar to that of peptide **14**, with the added methyl group point outward without any steric interference (Fig. 2.9b). Therefore, as in the case of **14**, the non-helical structure of **10a** is also stable. The peptide **2a** with R = Ph also has the similar situation (Fig. 2.9c).

However, this non-helical structure will be significantly destabilized when a methyl substitution is placed at (R)-configuration, due to severe steric clash with the backbone O atom on the third residue (Fig. 2.9b). To relieve this clash, the tether chain will rotate to make the methyl group away from this O atom, as the Cluster 2 in Fig. 2.9c. However, the energy may increase due to the conformational change in the tether. Also, the methyl group becomes in contact with another backbone O atom, probably interfere with its solvation by water. Thus, the most populated structure cluster for peptides **10b** is α-helical, in which the R group is quite comfortable, as the representative structure of the Cluster 1 in Figure S5c. This simulated structure is almost identical to the crystal structure, with RMSD of backbone and Cβ only 0.3Å (Fig. 2.9c).

Furthermore, in agreement with the experiments, simulated helicity of **2b** with R = Ph is higher than that of **1b** and **10b** with R = CH$_3$ (Table 2.7). From the representative non-helical structure of **2b**, one C atom on the phenyl group has steric

**Table 2.7** The calculated helicity from REMD simulations of 5 peptides

| Peptide | Seq. | Config. | Substitution | Calc. helicity (%) |
|---|---|---|---|---|
| **1a** | AAA | (*S*)- | CH3 | 0.2 |
| **1b** | AAA | (*R*)- | CH3 | 29.6 |
| **2a** | AAA | (*S*)- | -Ph | 1.1 |
| **2b** | AAA | (*R*)- | -Ph | 56.0 |
| **10a** | AAI | (*S*)- | CH3 | 3.5 |
| **10b** | AAI | (*R*)- | CH3 | 28.3 |
| **14** | AAA | – | – | 10.1 |

clash (3.1 Å distance) with the backbone O atom of the fourth residue, leading to destabilization and lower population (Cluster 2 in Fig. 2.9d). This steric interference may also occur for other non-helical conformations (Fig. 2.11). Indeed, this steric repulsion will limit the rotation of the peptide plane, making the $\psi$ angle of the 4th residue and the $\phi$ angle of the 5th residue to favor the α-helical conformation (Fig. 2.12). On the other hand, the phenyl group is quite comfortable when the backbone adopts -helix structure (Cluster 1 in Fig. 2.9d). Therefore, peptide 9b has much stronger preference for -helical conformation, as shown in its , plot.

### 2.2.8 *Cell Permeability Study*

The influence of the peptides' conformations on their biochemical/biophysical properties remains unclear, largely due to the absence of methods for constructing peptides with minimal differences in chemical composition. Cell permeability is the major limitation for peptide therapeutics and is influenced by many aspects, including conformation [44–47]. Scrambling the positions of a few amino acids in a peptide could dramatically change its permeability and other biophysical properties. However, our strategy provides an ideal platform for specifically investigating the sole influence of conformational differences. First, peptide diastereomers **11a/11b** FITC-βA-[cyclo-CRARS$_5$(2-Ph)]-NH$_2$ and **12a/12b** FITC-βA-[cyclo-CRRRS$_5$(2-Ph)]-NH$_2$ (βA: beta alanine; FITC: fluorescein isothiocyanate) were synthesized and separated. As shown in Fig. 2.13, the helical diastereomers **11b** and **12b** could successfully penetrate HEK293T cells within 2 h while the countpart diastereomers were much less permeable (Fig. 2.13). This led us to consider if a helical conformation itself could make peptides permeable. Peptide diastereomers **13a/13b** FITC-βA-[cyclo-CAKAS$_5$(2-Ph)]-NH$_2$ were subsequently tested. Peptide **13b** showed enhanced helicity over peptide **13a** (Fig. 2.14a), however; while it outperformed peptide **13a** in penetration of the cell membrane, peptide **13b** only showed minimal penetrative efficacy (Fig. 2.14b, c). These results suggested that although helical conformation itself may not guarantee peptides' permeability, it is a determining factor for the peptides' permeability. To date, this method excluded any peptide composition perturbation and provided an ideal platform to study the solely conformational influence on a peptide's permeability and other biophysical properties.

### 2.2.9 *Bioactive Peptide Construction*

The structural elucidation of estrogen receptor alpha (ERα) and mammal double minute 2 (MDM2) with their constrained peptide ligands clearly showed the interaction between the protein targets and the ligand tethers, mostly in the flat hydrophobic region surrounding the target ligand binding site [31, 32, 41]. Based on these results, we chose these two model targets to study the influences of peptide helicity and

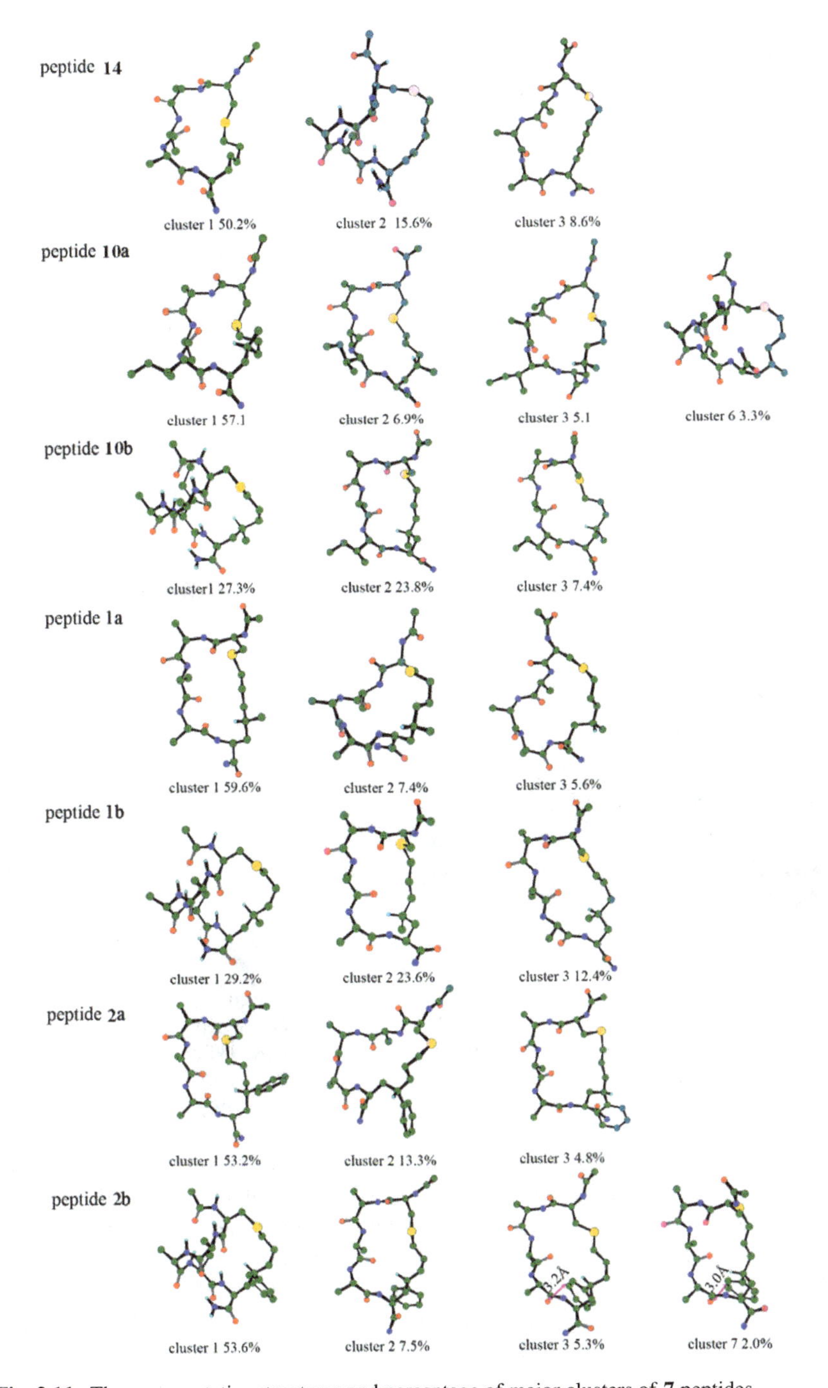


**Fig. 2.11** The representative structures and percentage of major clusters of **7** peptides

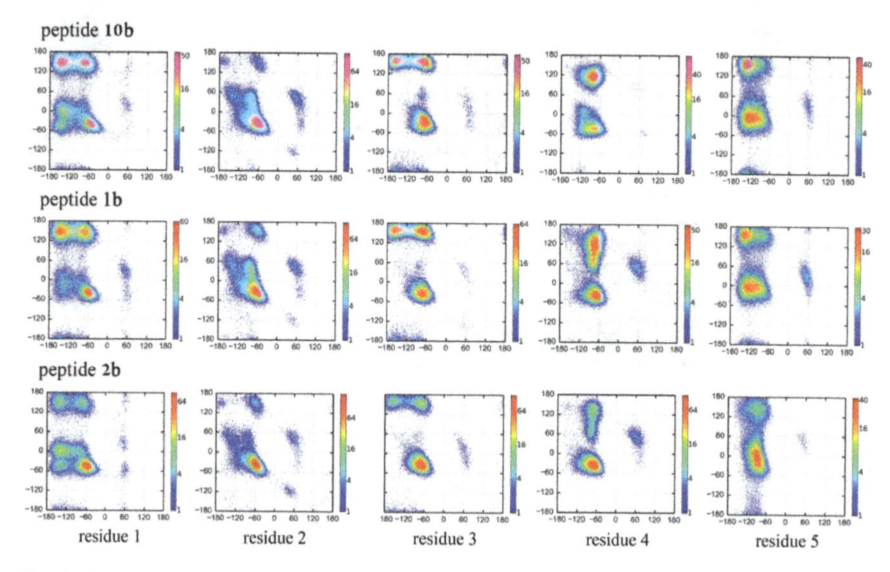


**Fig. 2.12** Ramachandran ($\phi$, $\psi$) plots of each residue from peptide **10b**, **1b**, and **2b** in simulation

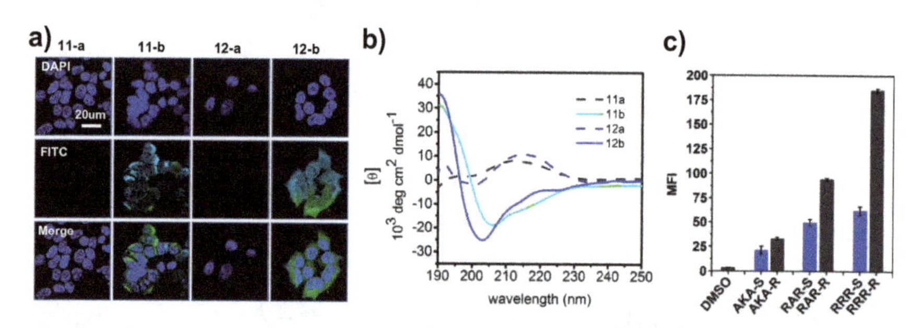


**Fig. 2.13 Cell permeability of pentapeptide diastereomers** . (**a**) Fluorescent confocal microscopy images of HEK293T cells incubated with FITC-labeled peptides **11a/b** and **12a/b** (5 μM) at 37°C for 2 h (blue (DAPI), green (FITC)). (**b**) CD spectra of peptides **11a/b** and 12a/b at 20°C in 50% TFE buffer. (**c**) Flow cytometry measurements of HEK293T cells with peptide **11a/b**, **12a/b**, and **13a/b** (5 μM) at 37°C for 2 h

substitution groups at the tether chiral center on the binding affinity for the peptides' target. **ER-1 a/b** and **ER-2 a/b** were synthesized based on their reported sequences (Scheme 2.5), which contain a methyl or phenyl group at the chiral center, respectively. **ER-1b** and **ER-2b** showed a significant increase in helicity compared to **ER-1a** and **ER-2a** (Fig. 2.15a). The binding affinity of **ER-1b** (~ 1 nM) and **ER-2b** (~ 69 nM) is much better than **ER-1a** (not determined) and **ER-2a** (> 600 nm) (Fig. 2.15b, c). Interestingly, **ER-1b** showed a significantly enhanced binding affinity compared to all previously reported ER-α peptide ligands, which may be caused by the additional interaction contributed by the methyl group at the stereo-center in the tether with the ERα protein. Peptides **PDI-1 a/b** and **PDI-2 a/b** were also synthesized

a)
b)
c)
13a
13b
DAPI
FITC
Bright field
Merge
20mm
AKA-R
Control
AKA-S
# Cells
FITC
Wavelength(nm)
[θ] 10³ Xdeg.cm²dmol⁻¹

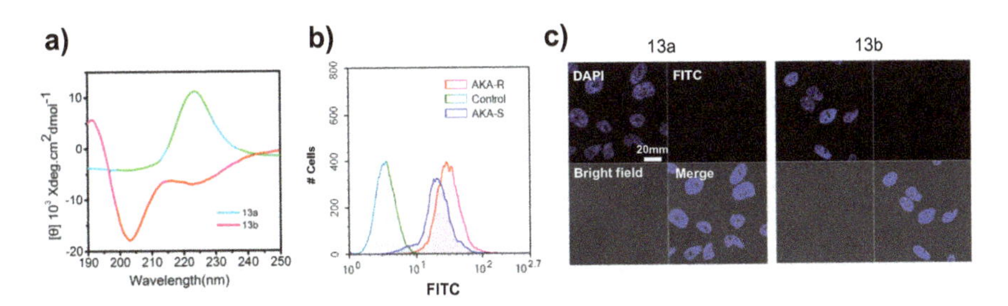

**Fig. 2.14** (**a**) The CD spectra of **13a/b** FITC-ßA-[cyclo-CAKAS$_5$(2-Ph)]-NH$_2$ in 50% TFE solution. (**b**) Flow cytometry measurement of **13a/b** after incubation with HEK293T cells at 37°C for 2 h. Peptide fluorescent confocal microscopy images of HEK293T cells incubated with FITC-labeled peptides **13a/b** FITC-ßA-[cyclo-CAKAS$_5$(2-Ph)]-NH$_2$ (5 μM) at 37°C for 1 h (DNA, blue (DAPI); peptides, green (FITC))

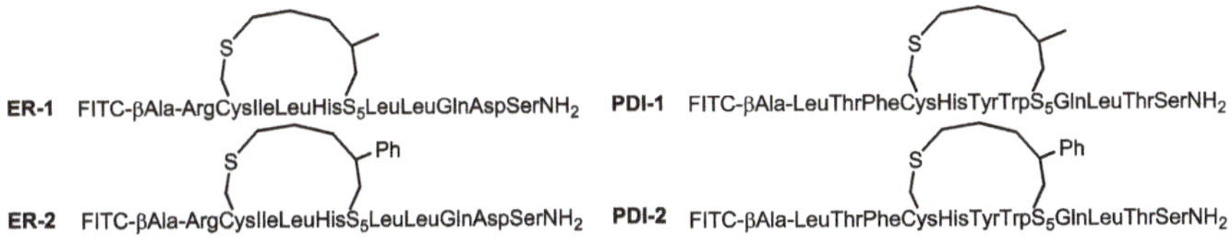

ER-1 FITC-βAla-ArgCysIleLeuHisS5LeuLeuGlnAspSerNH2
ER-2 FITC-βAla-ArgCysIleLeuHisS5LeuLeuGlnAspSerNH2
PDI-1 FITC-βAla-LeuThrPheCysHisTyrTrpS5GlnLeuThrSerNH2
PDI-2 FITC-βAla-LeuThrPheCysHisTyrTrpS5GlnLeuThrSerNH2
S
Ph

**Scheme 2.5 Target binding affinity of ERα & MDM2 with their peptide ligand diastereomers** . (A) Schematic presentation of **ER-1a/1b**, **ER-2a/2b**, **PDI-1a/1b**, and **PDI-2a/2b** structures

based on their reported sequences [54]. **PDI-1b** and **2b** showed a remarkable increase in helicity compared to **PDI-1a** and **2a** (Fig. 2.15d). They also showed significantly better binding affinities than **PDI-1a** and **2a** (Fig. 2.15e, f). However, **PDI-2b** showed unfavorable binding (~ 504 nM) compared to **PDI-1b** (~ 165 nM), which may cause by the steric hindrance posed by the bulky phenyl group and MDM2. These results constitute the first direct evidence of a direct relationship between helical enhancement and peptide ligand/protein target binding. Moreover, these results suggest that the substitution group also interacts directly with the binding groove, providing a valuable modification site for future applications such as fragment-based peptide ligand design.

Notably, cellular uptake experiments using MCF-7 cells treated with 5 μM peptides (**ER-1a/1b**, **PDI-1a/1b**) revealed that **PDI-1b** and **ER-1b** show significantly higher uptakes than their **diastereomers** (Fig. 2.16). These results were further confirmed by flow cytometry measurement (Fig. 2.17a–d). The peptide still penetrated the cell membrane when incubated at 4°C or with the addition of sodium azide (Fig. 2.16b). This suggests that the permeability mechanism could partly involve transduction and could be explained by the hydrophobic tether produced by the hydrophobic substitution group as well as the cyclization. Efforts are currently underway to elucidate the details of cell permeability. The in vitro serum stability assay showed that the **PDI-Linear** peptide degraded in a few hours, while more than 70% of peptides **PDI-1b** and **PDI-2b** remained intact after 24 h (Fig. 2.17e). Notably,

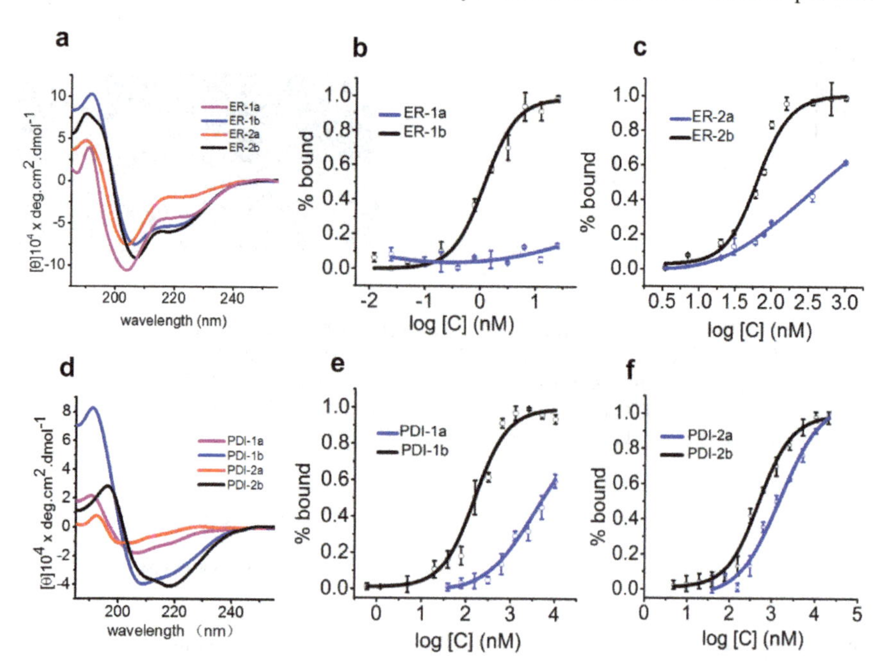


**Fig. 2.15** (A) CD spectra of ER peptides, measured in 20%TFE solution at 20°C (B, C). Binding of **ER-1a/1b** and **ER-2a/2b** with ERα, respectively. The binding affinities were measured using fluorescence polarization assays (FP) at 20°C. (D) CD spectra of **PDI** peptides, measured in 20% TFE solution at 20°C. (E, F) Binding of **PDI-1a/1b** and **PDI-2a/2b** with MDM2, respectively. The binding affinities were measured using fluorescence polarization assays (FP) at 20°C

the bulkier substitution group showed better proteolysis resistance. Thus, the chiral centre-induced helicity enhancement could be successfully translated into longer peptides with good binding affinity and intriguing cell permeability. More importantly, the additional substitution site may interact directly with the protein target and could be of interest for future medicinal development and other modifications.

## 2.3 Conclusion

In summary, a precisely-positioned in-tether carbon chiral center was found to be capable of modulating a peptide's helicity. This study provides an excellent platform for studying the relationship between a peptide's conformation and their biochemical/biophysical properties. We investigated the relationship between the helicities of the peptides and the location of the chiral center, the stereo configuration, the ring size and the size of the substitution group. The pentapeptide crystal structure and computational simulations further validate our results. Peptide diastereomers were also tested to examine the sole influence of conformation on cell permeability.

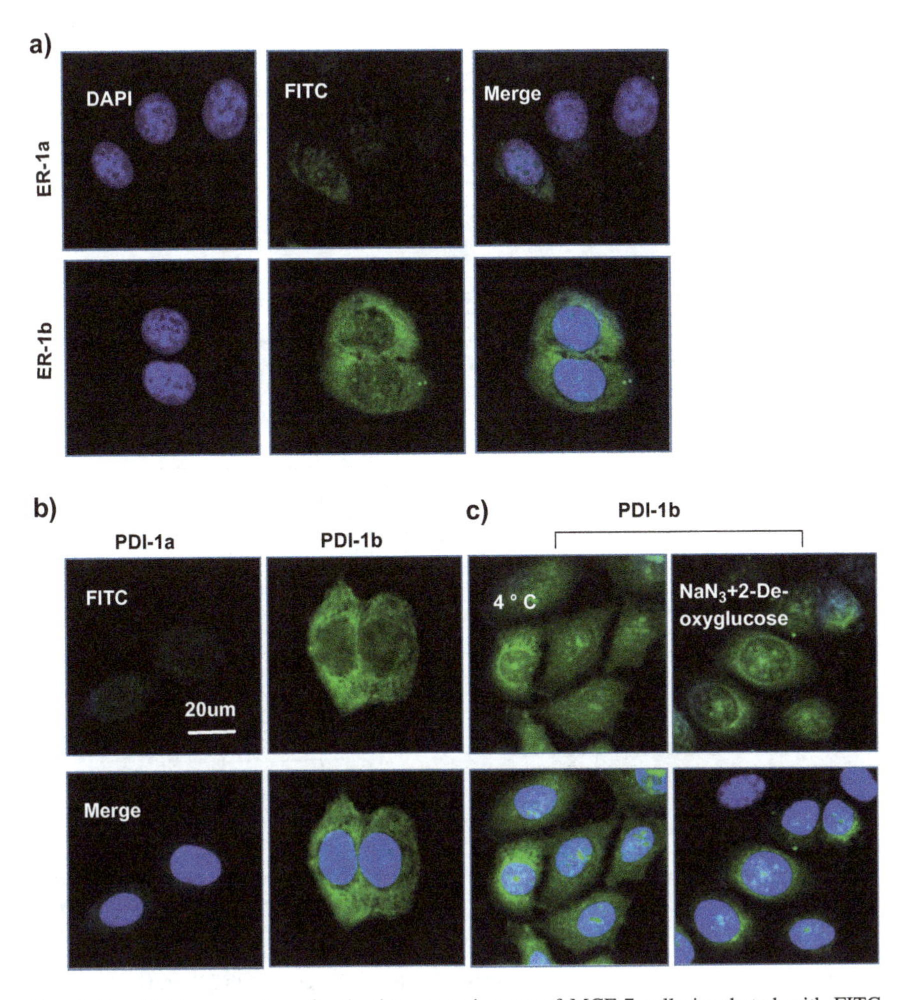


**Fig. 2.16** (A) Fluorescent confocal microscopy images of MCF-7 cells incubated with FITC-labeled peptides **ER-1a/1b** (5 μM) at 37°C. (B–D) Flow cytometry measurements of MCF-7 cells treated with **ER-1a/1b**, **PDI-1a/1b**, and **PDI-2a/2b (5** μM) at 37°C

Moreover, this concept was translated into constructing MDM2 and ERα peptide ligands, which show excellent α-helicity nucleation properties and dramatically enhanced binding affinities. More importantly, the significant differences in the permeability of the long peptide diastereomers clearly indicate the importance of increasing peptide helicity in the construction of constrained peptides. We unambiguously demonstrated that increasing the helicity of a constrained peptide could increase its permeability. In addition, the influence of the substitution groups at the chiral center on the peptides' binding affinity suggests that this chiral center could be utilized as an additional modification site away from the peptide backbone that

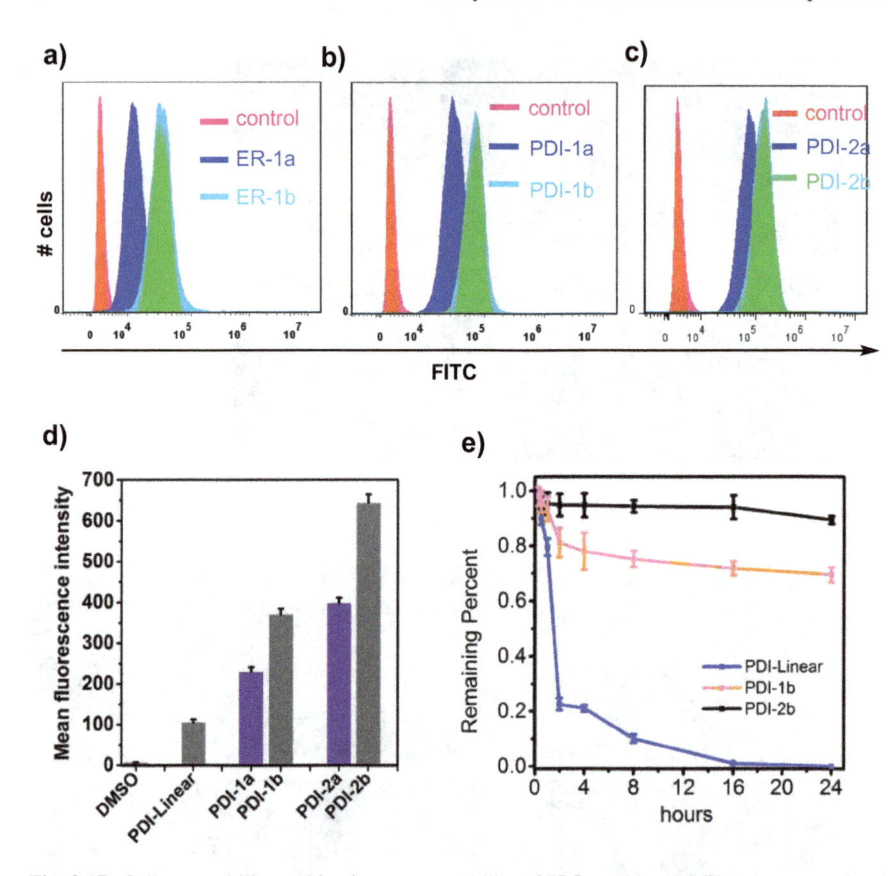


**Fig. 2.17** Cell permeability and in vitro serum stability of PDI peptides. (**a**) Fluorescent confocal microscopy images of MCF-7 cells incubated with FITC-labeled peptides **PDI-1a/1b** (5 μM) at 37°C. (**b**) Fluorescent confocal microscopy images of MCF-7 cells incubated with FITC-labeled peptides **PDI-1b** (5 μM) at 4°C or treated with $NaN_3$ + 2-Deoxyglucose for 1 h (DNA, blue (DAPI) peptides, green (FITC). (**c**) FACS measurements of MCF-7 cells treated with **ER-1a/1b**, **PDI-1a/1b**, and **PDI-2a/2b** (5 μM) at 37°C. For separated flowcytometry figures, see Figure S10. (**d**) In vitro serum digestion assay of **PDI-linear**, **PDI-1b**, and **PDI-2b** peptides

could be useful for various applications. Studies on the chiral center effect and further biological applications are currently underway and will be reported in due time.

## 2.4 Methods and Materials

### 2.4.1 Reagents and Equipment

All reagents including amino acids and resins were purchased from GL Biochem (Shanghai), Shanghai Hanhong Chemical Co., J&K Scientific or Energy Chemical and were used without further purifications. Unnatural amino acids were synthesized following reported procedures. NMP were purchased from Shenzhen Tenglong Logistics Co. and used without purification. All other solvents used were purchased from Cantotech Chemicals, Ltd. Anhydrous solvents were purchased from J&K Scientific. NMRs were measured on nuclear magnetic resonance (NMR) spectroscopy (Bruker AVANCE-III 300, 400 and 500). Variable-temperature NMR were acquired using Bruker AVANCE III 500 MHz and processed using Topspin 3.1 software package. Peptides were purified by HPLC (SHIMAZU Prominence LC-20AT) using reverse phase C18 column Grace Vydac protein and peptide C18 250 × 10 mm, flow rate 5 ml/min or grace smart C18 250 × 4.6 mm at flow rate 1 ml/min. Deionized $H_2O$ (containing 0.1% TFA) and pure acetonitrile were used as solvents in linear gradient elution. HPLC fractions containing product (screened by ESI) were combined and lyophilized. Molecular weights were measured on SHIMAZU-SPD2020; CD spectra were measured on Chirascan Circular Dichroism Spectrometer.

### 2.4.2 Peptide Purification and Characterization

Linear peptides were synthesized, and then characterized by HPLC and LC-MS. The photoreaction products were purified on the HPLC. In general, for peptides **1-10** and **s1-8a/b** the S/R diastereomers were separable with more than 5 min retention time differences. The purified peptides were then characterized by LC-MS. Mass spectra were obtained by ESI in positive ion mode. All LC-MS figures were attached at the end of supporting information.

### 2.4.3 CD Measurement

Peptides were dissolved in potassium phosphate solution (pH 7.0) or 20% TFE buffer to concentrations of 10–100 μM. The spectra were obtained on an Applied Photophysics Chirascan Circular Dichroism Spectrometer at 20°C using the following standard measurement parameters: wavelength, 185–260 nm; step resolution, 0.5 nm; speed, 20 nm/sec; accumulations, 10; response, 1 s; bandwidth, 1 nm; path 3 length, 0.1 cm. Every sample was scanned twice and the final CD spectrum was averaged and smoothed. The α-helical content of each peptide was calculated as reported previously. The final helical content presented as relative to peptide **2b**, as fixed the peptide **2b** as 100% helicity. Thermal disruption curves were acquired by monitoring the signal at 222 nm while the temperature was increased at 5 °C intervals with an equilibration time of 10 min between temperature increases. The data were fit to a two-state folding model using Origin Pro 9.0.

### 2.4.4 NMR Spectroscopy

NMR data were recorded on a Bruker AVANCE III 400 (or 500) MHz spectrometer. DMSO-$d^6$ was used for $^1$H NMR to characterize the peptides. NMR data were processed using Topspin 3.0. Temperature coefficients were used as tools to characterize the propensity for exchangeable protons to form intramolecular hydrogen bonds (IMHBs). For this experiment, the peptides were dissolved in 9:1 $H_2O$: $D_2O$. The protection of the IMHBs decreases the temperature dependence of the chemical shift of the exchangeable protons. This ultimately results in a smaller value of $\Delta\delta/T$ compared to non-IMHB donors. Generally, the cutoff value of $\Delta\delta/T$ for IMHBs is solvent dependent. In aqueous solution, values of $\Delta\delta/T$ more than -4 ppb/K usually indicate hydrogen bonding. 2D NMR data were collected on a Bruker Avance III 500 MHz spectrometer with a TXI probe. Watergate pulse sequence with gradients were used for water suppression in 1D and 2D 1H spectrum. 2D 1H-1H TOCSY and NOESY spectra were acquired with mixing time of 100 ms and 300 ms, respectively. The TOCSY and NOESY spectra were acquired with a width of 10 ppm and 13C spectra width of 100 ppm, and size of 1024 × 400 complex points. All the 2D NMR spectra were processed by Topspin ® to final 2048 × 1024 complex points, and analyzed by CCPNMR software. $^3J_{(NH\text{-}Ha)}$ couplings were measured from 1D-1H spectrum. Temperature dependence for amide NH chemical shifts was measured from 2D TOCSY spectra recorded at temperature ranges from 288 K to 313 K with 5 K interval. At each temperature, the sample was allowed to equilibrate for 15 min, and the chemical shifts were calibrated with standard 4,4-dimethyl-4-silapantane-1-sulfonic acid (DSS).

### 2.4.5 Crystallization and Data Collection

Peptide 10b. Ac-(cyclo-1,5)-[CAAIS$_5$(2-Me)]-$NH_2$ was dissolved in 50% $CH_3OH$ at 10 mg/ml and crystallized at 25°C using sitting drop vapor diffusion method against reservoir solution of 50% $CH_3OH$. Crystals are flash frozen in liquid nitrogen and cryo-protected by 30% glycerol in mother liquor. The crystals were screened and collected at 100 K by in-house X-ray diffraction system equipped with high-intensity sealed copper tube X-ray generator (Rigaku® MicroMax-002 +), an AFC11 goniometer, a Saturn 944 + CCD detector (Rigaku®), and an Oxford Cryo-system.

## 2.4.6 Structure Determination and Refinement

Data collection, integration, scaling, and empirical absorption correction were carried out in the Rigaku CrystalClear-2.054 program package. The structure was solved in 0.82Å resolution by direct method using the software of SIR201155 and well refined by Full-Matrix-Least-Squares against F2 by SHELXTL97 [55]. The non-hydrogen atoms were anisotropically refined and hydrogen atoms were placed at idealized positions and refined using the riding model. The absolute configuration was determined using the method of Flack [56]. The statistics of data collection and final refinement were shown in Table 2.4.

## 2.4.7 Computational Studies

### 2.4.7.1 Methods for Computational Studies

Simulation settings: All simulations were carried out using the Gromacs 4.5.4 [57]. Our recently developed residue-specific force field [58, 59] RSFF2 [60] was used to treat each peptide, except for the tether, which was described using the generalized Amber force field (GAFF) [60, 61] with the restrained electrostatic potential (RESP) [62] charges. Each starting structure was initially constructed using the HyperChem software, and solvated in a truncated octahedron box (30 Å in length) with 647-709 TIP3P water molecules (depending on the peptides). Energy minimization was carried out using steepest descent method. Then, the initial periodic box volume was equilibrated using a 3 ns MD simulation in an NPT ensemble near 300 K and 1 atm. Subsequently, the initial structures for REMD were obtained at regular intervals from a 30 ns NVT MD trajectory at 600 K. For each REMD simulation, 24 replicas were used with temperature range from 300 K to 600 K. By frequently exchanging the replicas of different temperatures, REMD can speed up the barrier crossing and achieve higher efficiency in conformational sampling [63].

The electrostatics were treated using the particle-mesh Ewald (PME) method [64] with a real-space cutoff of 0.9 nm and van der Waals interaction cutoff at 0.9 nm with the long-range dispersion correction for energy and pressure in all simulations. A velocity rescaling thermostat [65] with $\tau_T = 0.2$ ps and a Berendsen barostat [66] with $\tau_P = 0.5$ ps were used to maintain constant temperature and constant pressure (for NPT simulations), respectively. All bonds involving hydrogen were constrained using LINCS [67], and a time step of 2 fs was used. At the same time, the mass of water oxygen atom was reduced from 16 to 2 amu to increase the sampling efficiency [68] without altering the thermodynamics equilibrium properties. 24 replicas of the system were simulated simultaneously at temperatures from 300 to 600 K. The intermediate temperatures were chosen following a recent study [69] to obtain uniform exchange rate, and exchanges were attempted between neighboring replicas every 1.0 ps. The

REMD simulations were carried out for 200 ns, achieving a total simulation time of 4.8 μs for each peptide.

#### 2.4.7.2 Clustering Analysis, Calculated Helicity and Ramachandran Plots

Clustering analysis was performed on 10,000 snapshots sampled at 300 K, using the 'gromos' [70] method with 1Å cutoff based on RMSD of backbone and Cβ atoms. The representative structures and population of major clusters of 7 peptides were shown in Fig. 2.11. Calculated helicity (helical content) was defined as the percentage of the structures with all residues in the right-handed-helix conformation ($-160° < \phi < -40°$, $-80° < \psi < 40°$). The calculated helicity from REMD simulations of the five peptides is shown in Table 2.7.

For each peptide, the Ramachandran plot was drawn using the $\phi$, $\psi$ values of all five residues together, from 10,000 snapshots saved during the simulation. To draw the plot, the whole $360° \times 360°$ $\phi$, $\psi$ space is divided into $300 \times 300$ small bins, and the relative probability of each bin was calculated from the counts. A logarithmic color scale is used to visualize the probability distribution, and the bins not sampled are left in white.

### *2.4.8 Molecular Cloning, Protein Expression, and Purification*

(a) Cloning, Expression, and Purification of ER- LBD4. Human ER- LBD 301-553 was cloned into pET23b via NdeI and XhoI generating untagged constructs. Expression was carried out in *E. coli* BL21 (DE3) without IPTG induce. Cultures were grown in 2YT medium at 37°C to $OD_{600}$ of 0.8 and then transferred to 20°C for 18 h. Cells were harvested by centrifugation and flash frozen. Harvested cells were lysed via sonication in 100 mM lysis buffer (Tris-Cl pH 8.1, 300 mM KCl, 5 mM EDTA, 4 mM DTT, and 1 mmol/L PMSF). Cell debris were removed by centrifugation and the supernatant ran over a 1 mL estradiol affinity column (PDI technology) and the column was then eluted with elution buffer (100 μM estradiol, 20 mM Tris pH 8.1, and 0.25 M NaSCN). High molecular weight species and excess salts were removed on a Superdex 200 column equilibrated in buffer of 50 mM Tris pH 7.4, 150 mM NaCl, 10% glycerol, and 1 mM DTT.

Primer sequence:

ER-alpha-NdeI-301: GTGTACACATATGtctaagaagaacagcctggccttgt
ER-alpha-XhoI-553: Ccctcgagttaagtgggcgcatgtaggcggt

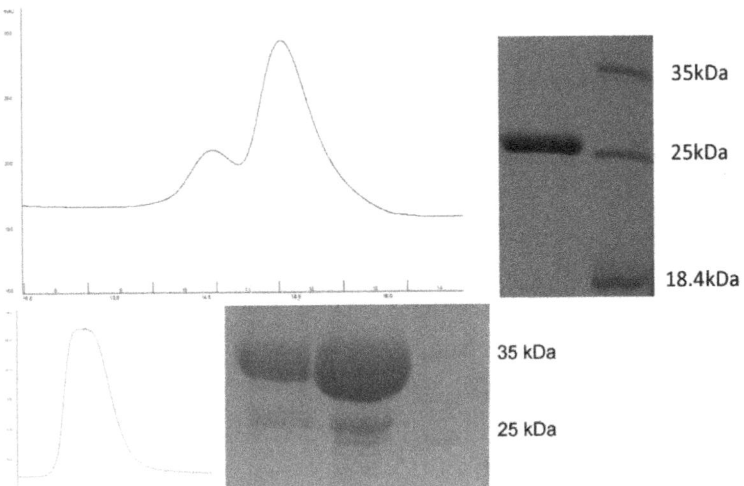


(b) Cloning, Expression, and Purification of MDM2. Human MDM2 LBD residues 25-117 was cloned into pGEX-4t-1 via EcoRI and XhoI generating GST-tagged constructs. Expression was carried out in *E. coli* BL21 (DE3) with expression being induced with 0.1 mM IPTG. Cultures were grown in LB medium at 37°C to an $OD_{600}$ of 0.6 before being transferred to 18°C for 24 h. Cells were harvested by centrifugation and flash frozen. Harvested cells were lysed by sonication in lysis buffer (20 mM Tris-Cl pH 7.9, 500 mM NaCl). Cell debris was removed by centrifugation and the supernatant was purified on a 5 mL GST affinity column (GE healthcare) and eluted with elution buffer (10 mM GSH in 20 mM Tris-Cl pH 7.9, 500 mM NaCl). The protein was further purified with Superdex 200 column equilibrated in 20 mM Tris-Cl pH 7.9, 500 mM NaCl, 1 mM DTT.

Primer sequence:

MDM2-EcoRI-25: CCGGAATTCGAGACCCTGGTTAGACCAAA
MDM2-XhoI-117: GTAGGCACTCGAGTCAGTCCGATGATTCCT

### 2.4.9 *Fluorescence Polarization*

a. ER-α/ER-1 peptide Fluorescence polarization experiments were performed in 96-well plates (Perkin Elmer Optiplate-96F) on plate reader (Perkin Elmer, Envision, 2104 multilabel reader). Concentrations of the peptides were determined by 495 nm absorption of FITC. Purified ER- LBD (at increasing concentrations, 20 μL) and fluorescein-labeled peptides (10 nM, 80 μL) in assay buffer (10 μM 17-β-estradiol, 20 mM Tris-HCl pH 8.0, 25 mM NaCl, 10% glycerol, 10 μM beta-estradiol, and 1 mM TCEP) were mixed and incubated at 4°C for 1 h in the dark. The fluorescence polarization of the labeled peptides was measured at 16°C with excitation at 485 nm and emission at 520 nm and then plotted against the concentrations of the ER- LBD. The data points were fitted by Origin pro 9.0.

b. $MDM_2$/PDI peptide FITC-labeled peptides (10–20 nM) were incubated with $HDM_2$ 17-125 in binding assay buffer (140 mM NaCl, 50 mM, Tris pH 8.0) at room temperature for 1 h. Fluorescence polarization experiments were performed in 96-well plates (Perkin Elmer Optiplate-96F) on plate reader (Perkin Elmer, Envision, 2104 multilabel reader). Concentrations of the peptides were determined by 494 nm absorption of FITC. $K_d$ values were determined by nonlinear regression analysis of dose response curves using Origin pro 9.0.

### *2.4.10 Cell Imaging*

HEK 293T cells (or Hela or MCF-7 cells) were cultured with DMEM with 10% FBS (v/v) in imaging dishes (50,000 cells/well) in 37°C, 5% $CO_2$ incubator for one day until they were about 80% adherent. Peptide were first dissolved in DMSO to make a 1 mM stock and then added to cells to a final concentration of 5 μM. The cells were incubated with peptides for 1 h at 37°C. After incubation, cells were washed 3 times with PBS and then fixed with 4% formaldehyde (Alfa Aesar, MA) in PBS for 10 min. They were then washed 3 times with PBS and stained with 1 μg/ml 4', 6-diamidino-2-phenylindole (DAPI) (Invitrogen, CA) in PBS for 5 min. Images of peptide localization in cells were taken on PerkinElmer confocal microscopy. Image processing was done using Volocity software package (Zeiss Imaging).

### *2.4.11 Flow Cytometry*

MCF-7 cells were grown in DMEM medium with 10% FBS (v/v) in imaging dishes (50,000 cells/well) in 37°C, 5% $CO_2$ incubator for two days (50,000 cells per well). Cells were treated with fluoresceinated peptides (5 μM) for up to 2 h at 37°C. After washing with media, the cells were exposed to trypsin (0.25%; Gibco) digestion (5 min, 37°C), washed with PBS, and resuspended in PBS. Cellular fluorescence was analyzed using a BD FACSCalibur flow cytometer (Becton Dickinson) and CellQuest Pro (or CFlow plus). The identical experiment was performed with 30 min pre-incubation of cells at 4°C followed by 4 h incubation with fluoresceinated peptides at 4°C to assess temperature dependence of fluorescent labeling.

### *2.4.12 Serum Stability*

The in vitro serum stability assay followed the procedure of previous literature [13]. Standard solution of PDI-linear (FITC-ßAla-LTFEHYWAQLTS-$NH_2$), PDI-1b (FITC-(cyclo-5,9)-[ßAla-LTFCHYWS5(2-Me)$_{(R)}$QLTS]-$NH_2$), and PDI-2b (FITC-(cyclo-5,9)-[ßAla-LTFCHYWS$_5$(2-Ph)$_{(R)}$QLTS]-$NH_2$) was prepared in water. Each peptide was added to the human serum (800 uL) and incubated at 37°C. Acetonitrile/water 3:1 was added to aliquots of serum at 0, 0.5, 1 h, 2 h, 4 h, 8 h, 16 h, and 24 h to precipitate serum proteins, which are removed by centrifugation. The standard supernatant was analyzed by LC/MS with a grace smart C18 250 × 4.6 mm column, using a 3% per minute linear gradient from 20% to 80% acetonitrile over 20 min. The amount of starting material left in each sample was quantified by LC/MS-based peak detection at 220 nm.

## Appendix

### *Mass Statistics Data for the Peptides and Structure of Unnatural Amino Acids*

See Table 2.8

### *1D and 2D NMR Spectra for 1b, 2b, and 10b*

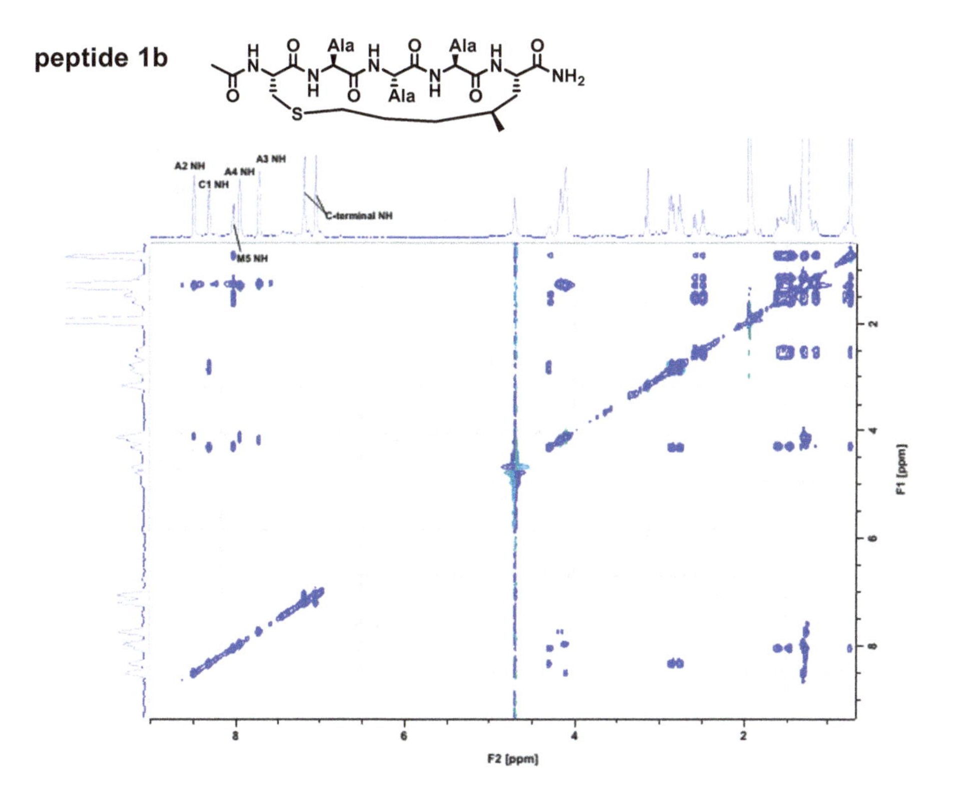

**Table 2.8** Calculated and founded m/z are presented as $[M + 1H]^{1+}/[M/2 + 1H]^{+}$

| Peptide | | Sequence | Calculated mass | Found mass |
|---|---|---|---|---|
| 1 | a | Ac-(cyclo-1,5)-[CAAAS$_5$(2-Me)(S)]-NH$_2$ | 514.26 | 515.30 |
| | b | Ac-(cyclo-1,5)-[CAAAS$_5$(2-Me)(R)]-NH$_2$ | 514.26 | 515.30 |
| 2 | a | Ac-(cyclo-1,5)-[CAAAS$_5$(2-Ph)(S)]-NH$_2$ | 576.27 | 577.30 |
| | b | Ac-(cyclo-1,5)-[CAAAS$_5$(2-Ph)(R)]-NH$_2$ | 576.27 | 577.30 |
| 3 | a | Ac-(cyclo-1,5)-[CAIAS$_5$(2-Me)(S)]-NH$_2$ | 556.3 | 557.40 |
| | b | Ac-(cyclo-1,5)-[CAIAS$_5$(2-Me)(R)]-NH$_2$ | 556.3 | 557.40 |
| 4 | a | Ac-(cyclo-1,5)-[CAEAS$_5$(2-Me)(S)]-NH$_2$ | 572.26 | 573.30 |
| | b | Ac-(cyclo-1,5)-[CAEAS$_5$(2-Me)(R)]-NH$_2$ | 572.26 | 573.30 |
| 5 | a | Ac-(cyclo-1,5)-[CASAS$_5$(2-Me)(S)]-NH$_2$ | 530.25 | 531.30 |
| | b | Ac-(cyclo-1,5)-[CASAS$_5$(2-Me)(R)]-NH$_2$ | 530.25 | 531.30 |
| 6 | a | Ac-(cyclo-1,5)-[CAQAS$_5$(2-Me)(S)]-NH$_2$ | 571.28 | 572.40 |
| | b | Ac-(cyclo-1,5)-[CAQAS$_5$(2-Me)(R)]-NH$_2$ | 571.28 | 572.40 |
| 7 | a | Ac-(cyclo-1,5)-[CAFAS$_5$(2-Me)(S)]-NH$_2$ | 590.29 | 591.50 |
| | b | Ac-(cyclo-1,5)-[CAFAS$_5$(2-Me)(R)]-NH$_2$ | 590.29 | 591.50 |
| 8 | a | Ac-(cyclo-1,5)-[CAGAS$_5$(2-Me)(S)]-NH$_2$ | 500.24 | 501.30 |
| | b | Ac-(cyclo-1,5)-[CAGAS$_5$(2-Me)(R)]-NH$_2$ | 500.24 | 501.30 |
| 9 | a | Ac-(cyclo-1,5)-[CEAKS$_5$(2-Me)(S)]-NH$_2$ | 629.32 | 630.40 |
| | b | Ac-(cyclo-1,5)-[CEAKS$_5$(2-Me)(R)]-NH$_2$ | 629.32 | 630.40 |
| 10 | a | Ac-(cyclo-1,5)-[CAAIS$_5$(2-Me)(S)]-NH$_2$ | 556.3 | 557.50 |
| | b | Ac-(cyclo-1,5)-[CAAIS$_5$(2-Me)(R)]-NH$_2$ | 556.3 | 557.50 |
| 11 | a | FITC-(cyclo-2,6)-[ßAlaCRARS$_5$(2-Ph)(S)]-NH$_2$ | 1166.48 | 583.40 |
| | b | FITC-(cyclo-2,6)-[ßAlaCRARS$_5$(2-Ph)(R)]-NH$_2$ | 1166.48 | 583.50 |
| 12 | a | FITC-(cyclo-2,6)-[ßAlaCRRRS$_5$(2-Ph)(S)]-NH$_2$ | 1251.54 | 626.00 |
| | b | FITC-(cyclo-2,6)-[ßAlaCRRRS$_5$(2-Ph)(R)]-NH$_2$ | 1251.54 | 626.00 |
| 13 | a | FITC-(cyclo-2,6)-[ßAlaCAKAS$_5$(2-Ph)(S)]-NH$_2$ | 1053.41 | 526.90 |
| | b | FITC-(cyclo-2,6)-[ßAlaCAKAS$_5$(2-Ph)(R)]-NH$_2$ | 1053.41 | 526.90 |
| s1 | a | Ac-(cyclo-1,5)-[CAAAS$_3$(2-Me)(S)]-NH$_2$ | 486.23 | 487.30 |
| | b | Ac-(cyclo-1,5)-[CAAAS$_3$(2-Me)(R)]-NH$_2$ | 486.23 | 487.30 |
| s2 | a | Ac-(cyclo-1,5)-[hCAAAS$_3$(2-Me)(S)]-NH$_2$ | 500.24 | 501.30 |
| | b | Ac-(cyclo-1,5)-[hCAAAS$_3$(2-Me)(R)]-NH$_2$ | 500.24 | 501.30 |
| s3 | a | Ac-(cyclo-1,5)-[hCAAAS$_5$(2-Me)(S)]-NH$_2$ | 528.27 | 529.40 |
| | b | Ac-(cyclo-1,5)-[hCAAAS$_5$(2-Me)(R)]-NH$_2$ | 528.27 | 529.40 |
| s4 | a | Ac-(cyclo-1,5)-[CAAAS$_5$(3-Me)(S)]-NH$_2$ | 514.26 | 515.30 |
| | b | Ac-(cyclo-1,5)-[CAAAS$_5$(3-Me)(R)]-NH$_2$ | 514.26 | 515.30 |
| s5 | a | Ac-(cyclo-1,5)-[CAAAS$_5$(3-Ph)(S)]-NH$_2$ | 576.27 | 577.30 |

(continued)

**Table 2.8** (continued)

| Peptide | | Sequence | Calculated mass | Found mass |
|---|---|---|---|---|
| | b | Ac-(cyclo-1,5)-[CAAA$S_5$(3-Ph)(R)]-$NH_2$ | 576.27 | 577.30 |
| s6 | a | Ac-(cyclo-1,5)-[CAAA$S_5$(5-Me)(S)]-$NH_2$ | 514.26 | 515.30 |
| | b | Ac-(cyclo-1,5)-[CAAA$S_5$(5-Me)(R)]-$NH_2$ | 514.26 | 515.30 |
| s7 | a | Ac-(cyclo-1,5)-[$S_5$(2-Me)(S)AAAC]-$NH_2$ | 514.26 | 515.30 |
| | b | Ac-(cyclo-1,5)-[$S_5$(2-Me)(R)AAAC]-$NH_2$ | 514.26 | 515.30 |
| s8 | a | Ac-(cyclo-1,5)-[$S_5$(3-Ph)(S)AAAC]-$NH_2$ | 576.27 | 577.30 |
| | b | Ac-(cyclo-1,5)-[$S_5$(3-Ph)(R)AAAC]-NH2 | 576.27 | 577.30 |
| PDI-1 | a | FITC-(cyclo-5,9)-[ßAla-LTFCHYW$S_5$(2-Me)(S)QLTS]-$NH_2$ | 1998.85 | 999.5 |
| | b | FITC-(cyclo-5,9)-[ßAla-LTFCHYW$S_5$(2-Me)(R)QLTS]-$NH_2$ | 1998.85 | 999.5 |
| PDI-2 | a | FITC-(cyclo-5,9)-[ßAla-LTFCHYW$S_5$(2-Ph)(S)QLTS]-$NH_2$ | 2060.87 | 1030.4 |
| | b | FITC-(cyclo-5,9)-[ßAla-LTFCHYW$S_5$(2-Ph)(R)QLTS]-$NH_2$ | 2060.87 | 1030.4 |
| PDI-L | | FITC-ßAla-LTFEHYWAQLTS-$NH_2$ | 1956.82 | 979.42 |
| ER-1 | a | FITC-(cyclo-3,7)-[ßAla-RCILH$S_5$(2-Me)(S)LLQDS]-$NH_2$ | 1858.85 | 929.78 |
| | b | FITC-(cyclo-3,7)-[ßAla-RCILH$S_5$(2-Me)(R)LLQDS]-$NH_2$ | 1858.85 | 929.78 |
| ER-2 | a | FITC-(cyclo-3,7)-[ßAla-RCILH$S_5$(2-Ph)(S)LLQDS]-$NH_2$ | 1858.85 | 929.75 |
| | b | FITC-(cyclo-3,7)-[ßAla-RCILH$S_5$(2-Ph)(R)LLQDS]-$NH_2$ | 1858.85 | 929.75 |

**TOCSY spectrum of 1b (500 MHz in H2O with 10% D2O at 298 K)**

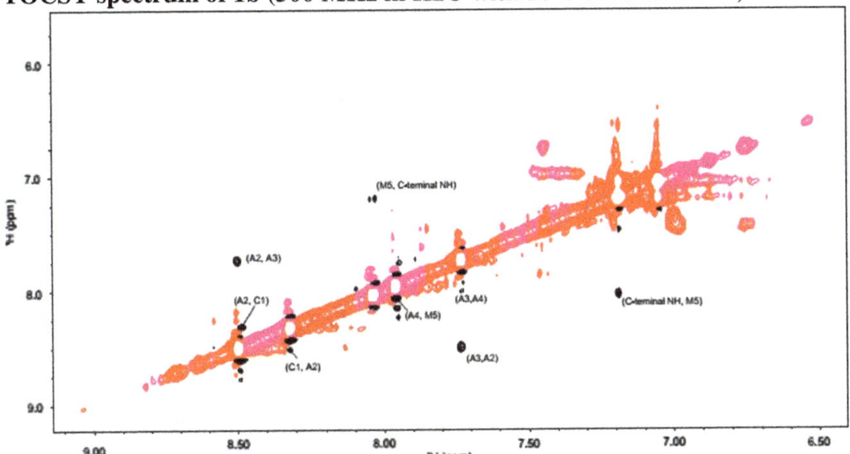


**NH region of NOESY spectrum of 1b (500 MHz in H2O with 10% D2O at 298 K)**

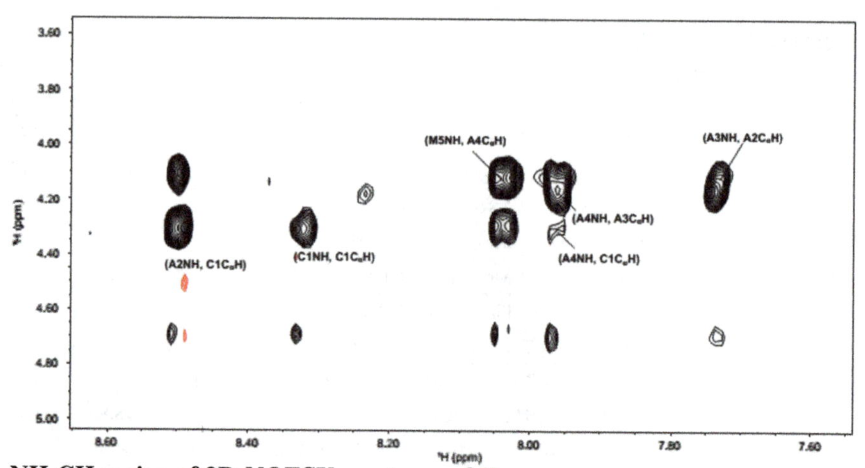


**NH-CH region of 2D-NOESY spectrum of 1b (at 500 MHz in $H_2O$ with 10% $D_2O$ at 298 K) asterisk indicates peak overlap)**

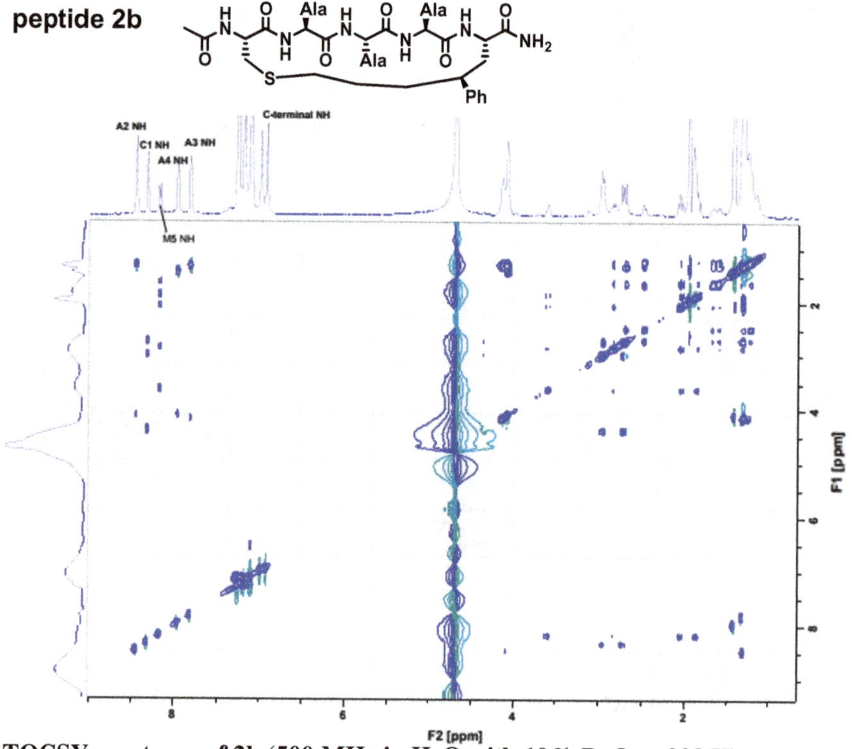


**TOCSY spectrum of 2b (500 MHz in $H_2O$ with 10% $D_2O$ at 298 K)**

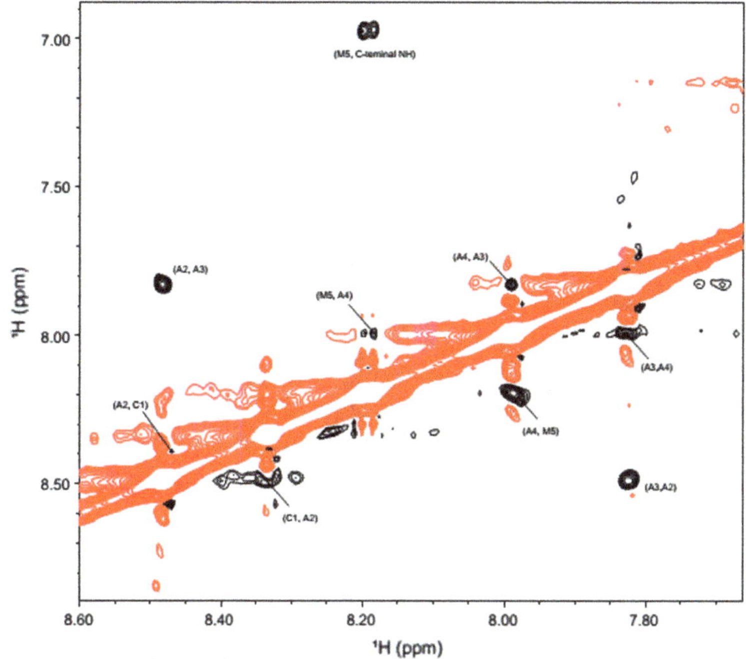


**NH region of NOESY spectrum of 2b (500 MHz in $H_2O$ with 10% $D_2O$ at 298 K)**

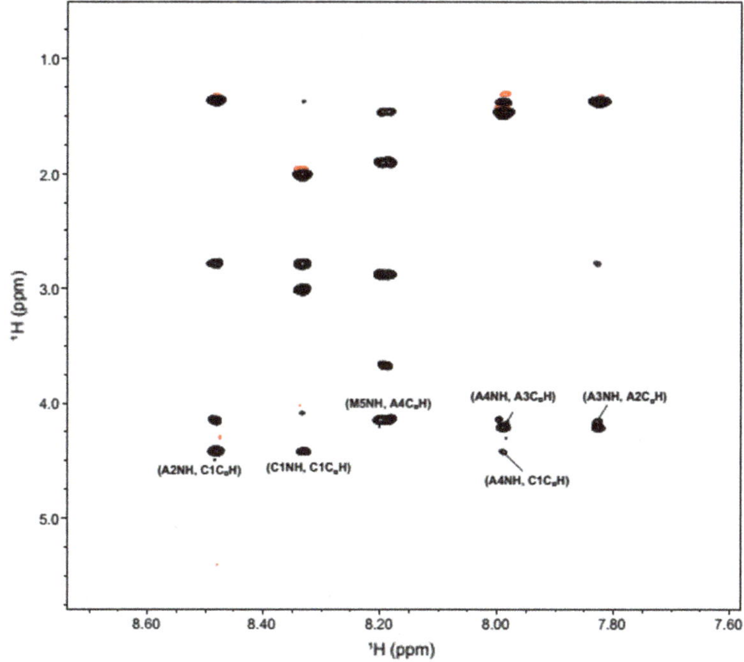


**NH-CH region of 2D-NOESY spectrum of 2b (at 500 MHz in $H_2O$ with 10% $D_2O$ at 298 K) asterisk indicates peak overlap)**

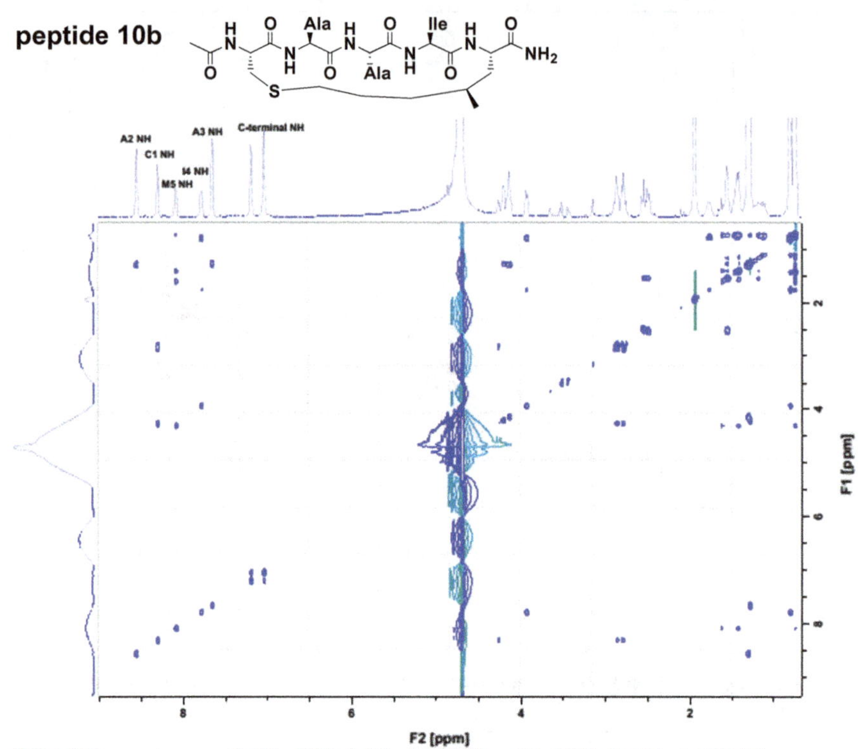


TOCSY spectrum of 10b (500 MHz in H2O with 10% D2O at 298 K)

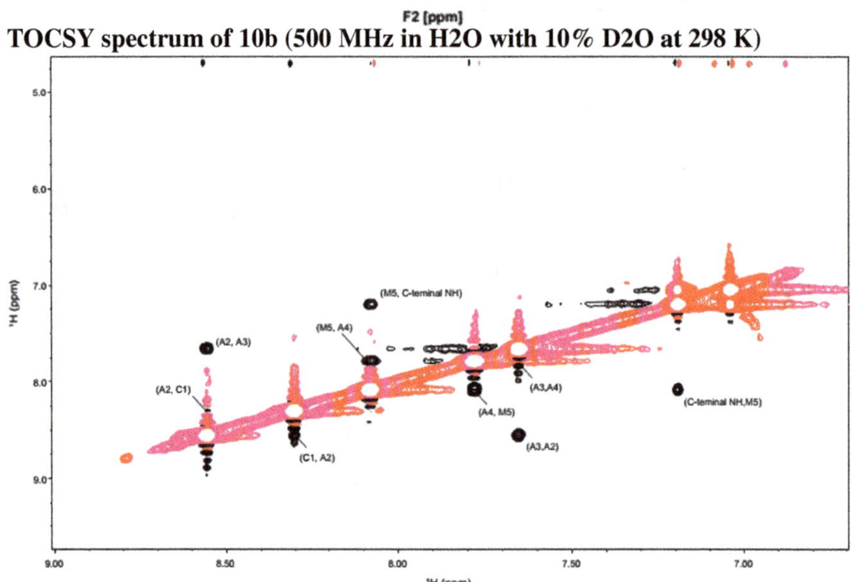


NH region of NOESY spectrum of 10b (500 MHz in $H_2O$ with 10% $D_2O$ at 298 K)

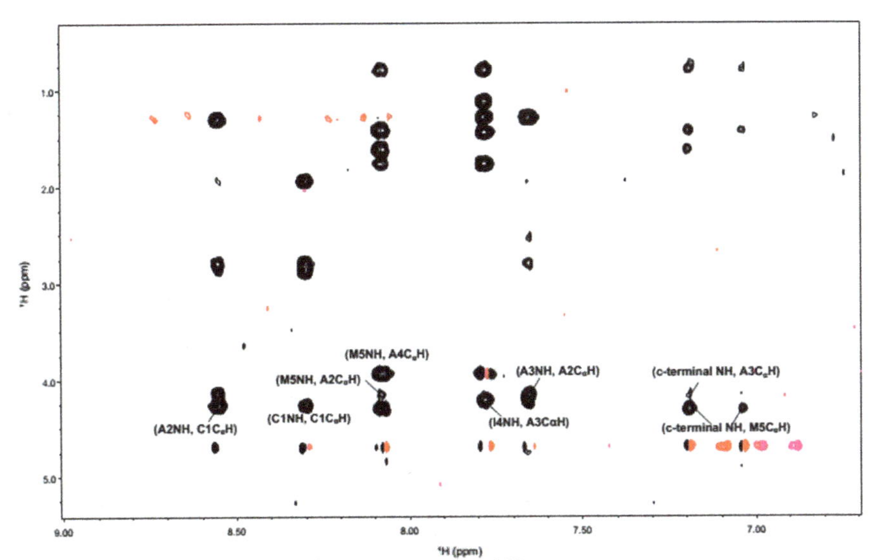


**NH-CH region of 2D-NOESY spectrum of 10b**
**(at 500 MHz in $H_2O$ with 10% $D_2O$ at 298 K) asterisk indicates peak overlap)**

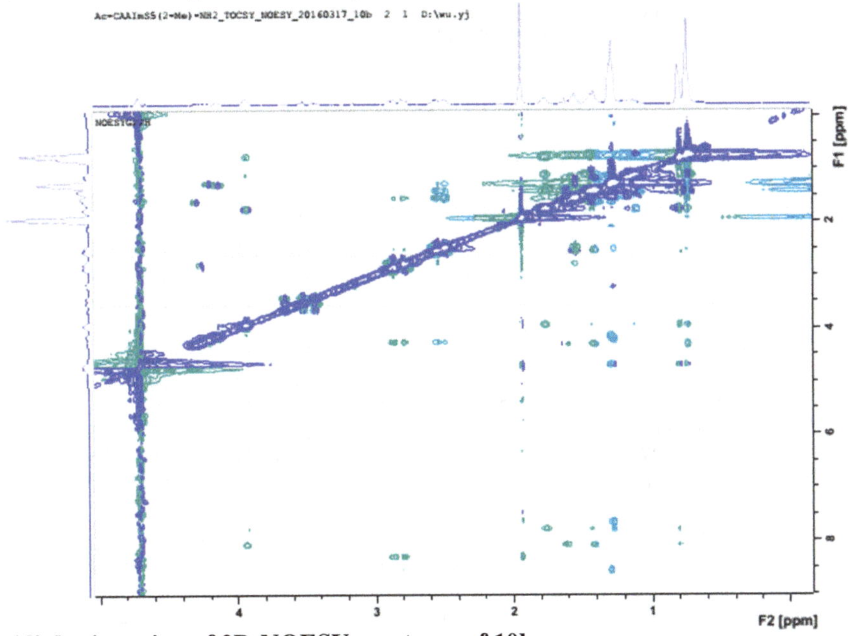


**Aliphatic region of 2D-NOESY spectrum of 10b**
**(at 500 MHz in $H_2O$ with 10% $D_2O$ at 298 K) asterisk indicates peak overlap)**

## References


1. Yin H, Hamilton AD (2005) Strategies for targeting protein-protein interactions with synthetic agents. Angew Chem Int Ed 44:4130–4163
2. Azzarito V, Long K, Murphy NS, Wilson AJ (2013) Inhibition of [alpha]-helix-mediated protein-protein interactions using designed molecules. Nat Chem 5:161–173
3. Milroy L-G, Grossmann TN, Hennig S, Brunsveld L, Ottmann C (2014) Modulators of protein-protein interactions. Chem Rev 114:4695–4748
4. Osapay G, Taylor JW (1990) Multicyclic polypeptide model compounds: 1. Synthesis of a tricyclic amphiphilic.alpha.-helical peptide using an oxime resin, segment-condensation approach. J Am Chem Soc 112:6046–6051
5. Phelan JC, Skelton NJ, Braisted AC, McDowell RS (1997) A general method for constraining short peptides to an α-helical conformation. J Am Chem Soc 119:455–460
6. Pease JHB, Storrs RW, Wemmer DE (1990) Folding and activity of hybrid sequence, disulfide-stabilized peptides. Proc Natl Acad Sci U S A 87:5643–5647
7. Jackson DY, King DS, Chmielewski J, Singh S, Schultz PG (1991) General approach to the synthesis of short.alpha.-helical peptides. J Am Chem Soc 113:9391–9392
8. Cabezas E, Satterthwait AC (1999) The hydrogen bond mimic approach: solid-phase synthesis of a peptide stabilized as an α-helix with a hydrazone link. J Am Chem Soc 121:3862–3875
9. Kumita JR, Smart OS, Woolley GA (2000) Photo-control of helix content in a short peptide. Proc Natl Acad Sci U S A 97:3803–3808
10. Schafmeister CE, Po J, Verdine GL (2000) An all-hydrocarbon cross-linking system for enhancing the helicity and metabolic stability of peptides. J Am Chem Soc 122:5891–5892
11. Leduc AM, Trent JO, Wittliff JL, Bramlett KS, Briggs SL, Chirgadze NY, Wang Y, Burris TP, Spatola AF (2003) Helix-stabilized cyclic peptides as selective inhibitors of steroid receptor-coactivator interactions. Proc Natl Acad Sci U S A 100:11273–11278
12. Galande AK, Bramlett KS, Burris TP, Wittliff JL, Spatola AF (2004) Thioether side chain cyclization for helical peptide formation: inhibitors of estrogen receptor–coactivator interactions. J Pept Res 63:297–302
13. Shepherd NE, Hoang HN, Abbenante G, Fairlie DP (2005) Single turn peptide alpha helices with exceptional stability in water. J Am Chem Soc 127:2974–2983
14. Cantel S, Le Chevalier Isaad A, Scrima M, Levy JJ, DiMarchi RD, Rovero P, Halperin JA, D'Ursi AM, Papini AM, Chorev M (2008) Synthesis and conformational analysis of a cyclic peptide obtained via i to i + 4 intramolecular side-chain to side-chain azide – alkyne 1, 3-Dipolar cycloaddition. J Org Chem 73:5663–5674
15. Muppidi A, Wang Z, Li X, Chen J, Lin Q (2011) Achieving cell penetration with distance-matching cysteine cross-linkers: a facile route to cell-permeable peptide dual inhibitors of Mdm2/Mdmx. Chem Commun 47:9396–9398
16. Lau YH, de Andrade P, Quah ST, Rossmann M, Laraia L, Skold N, Sum TJ, Rowling PJE, Joseph TL, Verma C, Hyvonen M, Itzhaki LS, Venkitaraman AR, Brown CJ, Lane DP, Spring DR (2014) Functionalised staple linkages for modulating the cellular activity of stapled peptides. Chem Sci 5:1804–1809
17. Haney CM, Horne WS (2014) Dynamic covalent side-chain cross-links via intermolecular oxime or hydrazone formation from bifunctional peptides and simple organic linkers. J Pept Sci 20:108–114
18. Zou Y, Spokoyny AM, Zhang C, Simon MD, Yu H, Lin Y-S, Pentelute BL (2014) Convergent diversity-oriented side-chain macrocyclization scan for unprotected polypeptides. Org Biomol Chem 12:566–573
19. Hilinski GJ, Kim Y-W, Hong J, Kutchukian PS, Crenshaw CM, Berkovitch SS, Chang A, Ham S, Verdine GL (2014) Stitched α-helical peptides via bis ring-closing metathesis. J Am Chem Soc 136:12314–12322
20. Mazzier D, Peggion C, Toniolo C, Moretto A (2014) Enhancement of the helical content and stability induced in a linear oligopeptide by an i, i + 4 intramolecularly double stapled, overlapping, bicyclic 31,22,5—(E)ene Motif. Biopolymers 102:115–123

21. Andrews MJI, Tabor AB (1999) Forming stable helical peptides using natural and artificial amino acids. Tetrahedron 55:11711–11743
22. Henchey LK, Jochim AL, Arora PS (2008) Contemporary strategies for the stabilization of peptides in the alpha-helical conformation. Curr Opin Chem Biol 12:692–697
23. Hill TA, Shepherd NE, Diness F, Fairlie DP (2014) Constraining cyclic peptides to mimic protein structure motifs. Angew Chem Int Ed Engl 53:13020–13041
24. Walensky LD, Bird GH (2014) Hydrocarbon-stapled peptides: principles, practice, and progress. J Med Chem 57:6275–6288
25. Lau YH, de Andrade P, Wu Y, Spring DR (2015) Peptide stapling techniques based on different macrocyclisation chemistries. Chem Soc Rev 44:91–102
26. Bracken C, Gulyas J, Taylor JW, Baum J (1994) Synthesis and nuclear-magnetic-resonance structure determination of an alpha-helical, bicyclic, lactam-bridged hexapeptide. J Am Chem Soc 116:6431–6432
27. Blackwell HE, Grubbs RH (1998) Highly efficient synthesis of covalently cross-linked peptide helices by ring-closing metathesis. Angew Chem Int Ed 37:3281–3284
28. Walensky LD, Kung AL, Escher I, Malia TJ, Barbuto S, Wright RD, Wagner G, Verdine GL, Korsmeyer SJ (2004) Activation of apoptosis in vivo by a hydrocarbon-stapled bh3 helix. Science 305:1466–1470
29. Moellering RE, Cornejo M, Davis TN, Bianco CD, Aster JC, Blacklow SC, Kung AL, Gilliland DG, Verdine GL, Bradner JE (2009) Direct inhibition of the NOTCH transcription factor complex. Nature 462:182–188
30. Stewart ML, Fire E, Keating AE, Walensky LD (2010) The MCL-1 BH3 helix is an exclusive MCL-1 inhibitor and apoptosis sensitizer. Nat Chem Biol 6:595–601
31. Baek S, Kutchukian PS, Verdine GL, Huber R, Holak TA, Lee KW, Popowicz GM (2012) Structure of the stapled p53 peptide bound to Mdm2. J Am Chem Soc 134:103–106
32. Chang YS, Graves B, Guerlavais V, Tovar C, Packman K, To K-H, Olson KA, Kesavan K, Gangurde P, Mukherjee A, Baker T, Darlak K, Elkin C, Filipovic Z, Qureshi FZ, Cai H, Berry P, Feyfant E, Shi XE, Horstick J, Annis DA, Manning AM, Fotouhi N, Nash H, Vassilev LT, Sawyer TK (2013) Stapled alpha-helical peptide drug development: a potent dual inhibitor of MDM2 and MDMX for p53-dependent cancer therapy. Proc Natl Acad Sci U S A 110:E3445–E3454
33. Sinclair JKL, Denton EV, Schepartz A (2014) Inhibiting epidermal growth factor receptor at a distance. J Am Chem Soc 136:11232–11235
34. Chapman RN, Dimartino G, Arora PS (2004) A highly stable short α-helix constrained by a main-chain hydrogen-bond surrogate. J Am Chem Soc 126:12252–12253
35. Wang D, Liao W, Arora PS (2005) Enhanced metabolic stability and protein-binding properties of artificial alpha helices derived from a hydrogen-bond surrogate: application to Bcl-xL. Angew Chem Int Edit 44:6525–6529
36. Liu J, Wang D, Zheng Q, Lu M, Arora PS (2008) Atomic structure of a short α-helix stabilized by a main chain hydrogen-bond surrogate. J Am Chem Soc 130:4334–4337
37. Henchey LK, Kushal S, Dubey R, Chapman RN, Olenyuk BZ, Arora PS (2010) Inhibition of hypoxia inducible factor 1—transcription coactivator interaction by a hydrogen bond surrogate α-helix. J Am Chem Soc 132:941–943
38. Frost JR, Vitali F, Jacob NT, Brown MD, Fasan R (2013) Macrocyclization of organo-peptide hybrids through a dual bio-orthogonal ligation: insights from structure-reactivity studies. Chem Bio Chem 14:147–160
39. Jo H, Meinhardt N, Wu Y, Kulkarni S, Hu X, Low KE, Davies PL, DeGrado WF, Greenbaum DC (2012) Development of α-helical calpain probes by mimicking a natural protein-protein interaction. J Am Chem Soc 134:17704–17713
40. Muppidi A, Li XL, Chen JD, Lin Q (2011) Conjugation of spermine enhances cellular uptake of the stapled peptide-based inhibitors of p53-Mdm2 interaction. Bioorg Med Chem Lett 21:7412–7415
41. Phillips C, Roberts LR, Schade M, Bazin R, Bent A, Davies NL, Moore R, Pannifer AD, Pickford AR, Prior SH, Read CM, Scott A, Brown DG, Xu B, Irving SL (2011) Design and structure of stapled peptides binding to estrogen receptors. J Am Chem Soc 133:9696–9699

42. Assem N, Ferreira DJ, Wolan DW, Dawson PE (2015) Acetone-linked peptides: a convergent approach for peptide macrocyclization and labeling. Angew Chem Int Edit 54:8665–8668
43. Brown SP, Smith AB (2015) Peptide/protein stapling and unstapling: introduction of s-tetrazine, photochemical release, and regeneration of the peptide/protein. J Am Chem Soc 137:4034–4037
44. Boguslavsky V, Hruby VJ, O'Brien DF, Misicka A, Lipkowski AW (2003) Effect of peptide conformation on membrane permeability. J Pept Res 61:287–297
45. Rezai T, Yu B, Millhauser GL, Jacobson MP, Lokey RS (2006) Testing the conformational hypothesis of passive membrane permeability using synthetic cyclic peptide diastereomers. J Am Chem Soc 128:2510–2511
46. Rezai T, Bock JE, Zhou MV, Kalyanaraman C, Lokey RS, Jacobson MP (2006) Conformational flexibility, internal hydrogen bonding, and passive membrane permeability: successful in Silico prediction of the relative permeabilities of cyclic peptides. J Am Chem Soc 128:14073–14080
47. Chu Q, Moellering RE, Hilinski GJ, Kim YW, Grossmann TN, Yeh JTH, Verdine GL (2015) Towards understanding cell penetration by stapled peptides. Med chem comm 6:111–119
48. Belokon YN, Tararov VI, Maleev VI, Savel'eva TF, Ryzhov MG (1998) Improved procedures for the synthesis of (S)-2-[N-(N′-benzylprolyl)amino] benzophenone (BPB) and Ni(II) complexes of Schiff's bases derived from BPB and amino acids. Tetrahedron Asymmetry 9:4249–4252
49. Tang X, Soloshonok VA, Hruby VJ (2000) Convenient, asymmetric synthesis of enantiomerically pure 2′,6′-dimethyltyrosine (DMT) via alkylation of chiral equivalent of nucleophilic glycine. Tetrahedron Asymmetry 11:2917–2925
50. Qiu W, Soloshonok VA, Cai C, Tang X, Hruby VJ (2000) Convenient, large-scale asymmetric synthesis of enantiomerically pure trans-cinnamylglycine and -α-alanine. Tetrahedron 56:2577–2582
51. Soloshonok VA, Tang X, Hruby VJ (2001) Large-scale asymmetric synthesis of novel sterically constrained 2′,6′-dimethyl- and α,2′,6′-trimethyltyrosine and -phenylalanine derivatives via alkylation of chiral equivalents of nucleophilic glycine and alanine. Tetrahedron 57:6375–6382
52. Aillard B, Robertson NS, Baldwin AR, Robins S, Jamieson AG (2014) Robust asymmetric synthesis of unnatural alkenyl amino acids for conformationally constrained [small alpha]-helix peptides. Org Biomol Chem 12:8775–8782
53. Wüthrich K, Billeter M, Braun W (1984) Polypeptide secondary structure determination by nuclear magnetic resonance observation of short proton-proton distances. J Mol Biol 180:715–740
54. Hu B, Gilkes DM, Chen J (2007) Efficient p53 activation and apoptosis by simultaneous disruption of binding to MDM2 and MDMX. Cancer Res 67:8810–8817
55. Pflugrath J (1999) The finer things in X-ray diffraction data collection. Acta Crystallogr Sect D 55:1718–1725
56. Sheldrick GM (2010) Experimental phasing with SHELXC/D/E: combining chain tracing with density modification. Acta Crystallogr Sect D-Biol Crystallogr 66:479–485
57. Pronk S, Páll S, Schulz R, Larsson P, Bjelkmar P, Apostolov R, Shirts MR, Smith JC, Kasson PM, van der Spoel D, Hess B, Lindahl E (2013) GROMACS 4.5: A high-throughput and highly parallel open source molecular simulation toolkit. Bioinformatics 29:845–854
58. Jiang F, Wu Y-D (2014) Folding of fourteen small proteins with a residue-specific force field and replica-exchange molecular dynamics. J Am Chem Soc 136:9536–9539
59. Jiang F, Zhou C-Y, Wu Y-D (2014) Residue-specific force field based on the protein coil library—RSFF1: modification of OPLS-AA/L. J Phys Chem B 118:6983–6998
60. Zhou C-Y, Jiang F, Wu Y-D (2015) Residue-specific force field based on protein coil library—RSFF2: modification of AMBER ff99SB. J Phys Chem B 119:1035–1047
61. Wang J, Wolf RM, Caldwell JW, Kollman PA, Case DA (2004) Development and testing of a general amber force field. J Comput Chem 25:1157–1174
62. Bayly C, Cieplak P, Cornell W, Kollman P (1993) A well-behaved electrostatic potential based method using charge restraints for deriving atomic charges: the RESP model. J Phys Chem 97:10269–10280

63. Sugita Y, Okamoto Y (1999) Replica-exchange molecular dynamics method for protein folding. Chem Phys Lett 314:141–151
64. Darden T, York D, Pedersen L (1993) Particle mesh Ewald: an N-log(N) method for Ewald sums in large systems. J Chem Phys 98:10089–10092
65. Bussi G, Donadio D, Parrinello M (2007) Canonical sampling through velocity rescaling. J Chem Phys 126:014101
66. Berendsen HJC, Postma JPM, Gunsteren WFv, DiNola A, Haak JR (1984) Molecular dynamics with coupling to an external bath. J Chem Phys 81:3684–3690
67. Hess B, Bekker H, Berendsen HJC, Fraaije JGEM (1997) LINCS: a linear constraint solver for molecular simulations. J Comput Chem 18:1463–1472
68. Lin IC, Tuckerman ME (2010) Enhanced conformational sampling of peptides via reduced side-chain and solvent masses. J Phys Chem B 114:15935–15940
69. Prakash MK, Barducci A, Parrinello M (2011) Replica temperatures for uniform exchange and efficient roundtrip times in explicit solvent parallel tempering simulations. J Chem Theory Comput 7:2025–2027
70. Daura X, Gademann K, Jaun B, Seebach D, van Gunsteren WF, Mark AE (1999) Peptide folding: when simulation meets experiment. Angew Chem Int Ed 38:236–240

# Chapter 3
# In-Tether Chiral Center Induced Helical Peptide Modulators Target p53-MDM2/MDMX and Inhibit Tumor Growth in Cancer Stem Cell

## 3.1 Introduction

Cancer is one of the most formidable diseases to combat. The cancer stem cell (CSC) hypothesis provides a compelling cellular mechanism to account for the therapeutic refractoriness and dormancy in cancer development. CSCs are a subset of stem-like cells that exhibit a unique spectrum of biologic, biochemical, and molecular features and possess the ability to efficiently propagate the bulk of tumors [1–3]. Their unlimited self-renewal and multipotency make CSCs more resistant to conventional and targeted therapies [1, 4]. CSCs are believed to be the primary cause of tumor recurrence and metastases [1, 2, 4, 5]. In clinical practice, CSCs need to be eradicated for long-term disease-free survival [2, 5].

Conventional anticancer agents predominantly target tumor bulk populations instead of CSCs. Increasing efforts have been directed toward the development of CSC therapy, and new approaches such as nanomedicine have been employed for the development of anti-CSC drugs [5–7]. In addition to self-renewal and pluripotency, the high-level expression of the adenosine triphosphate–binding cassette (ABC) transporters on CSC membranes enables CSC self-protection and constitutive drug resistance against anticancer drugs [8, 9]. Drug resistance and issues in drug delivery are the main obstacles in developing CSC therapy, and progress is compromised by the lack of compounds with suitable biological functions and pharmacological properties [6, 7] (Fig. 3.1).

Peptide stabilization is a technique for constraining short peptides into a fixed secondary conformation, typically an α- helix [10, 11]. As protein-protein interactions (PPIs) were previously thought to be 'undruggable' by small molecules due to their limited interacting surface area, stabilized peptides have become a promising drug modality to target PPIs [12]. We precisely added a chiral center into the peptide tether to constrain peptides into an α-helical conformation as shown in Fig. 3.2. Compared to the S diastereomers, the R diastereomer peptide showed significantly enhanced helical content, cellular uptake, and target binding affinity [13]. One major


K. Hu, *Development of In-Tether Carbon Chiral Center-Induced Helical Peptide*, Springer Theses,
https://doi.org/10.1007/978-981-33-6613-8_3

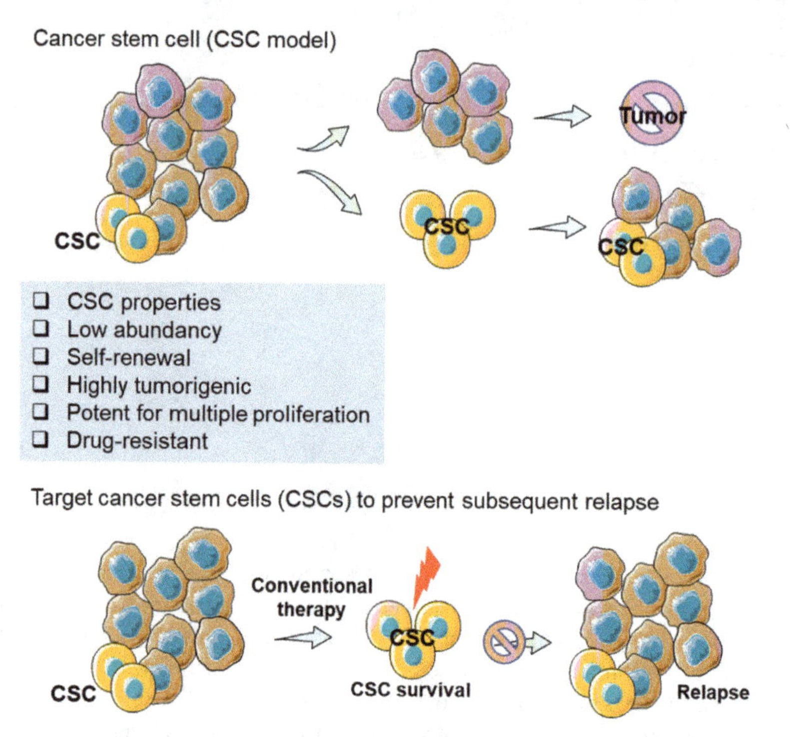


**Fig. 3.1** The concept of the cancer stem cell (CSC). Tumor cells are heterogeneous which contain a majority of cells are non/poorly tumorigenic, and a small subset of CSCs. The CSCs can be functionally distinguished from other populations by their ability to reconstitute a differentiated tumor upon transplantation into an immunocompromised mouse. Based on this model, CSC specific therapies are proposed in combination with conventional chemotherapies to kill both CSC and other differentiated populations and prevent subsequent relapse

advantage of our chirality-induced helicity (CIH) strategy is that we are able to fine-tune the peptides' biophysical properties, including target binding affinity and cellular uptake, by switching the substitution group at the chiral center. While stabilized peptides have previously been used to target various PPIs [10, 11, 14–16], to our knowledge, there has been no report of using stabilized peptides to regulate PPIs in CSCs or stem-like cancer cells. In our initial trials, we found that short peptides constructed via our strategy could successfully penetrate CSCs, and we hypothesized that stabilized peptide modulators could be developed to interrupt PPIs in CSCs and inhibit CSC growth and differentiation.

The human transcription factor protein p53 plays a pivotal role in protecting cells from malignant transformation by inducing cell-cycle arrest and apoptosis in response to DNA damage and cellular stress [17–19]. The ubiquitin E3 ligase MDM2 is the key factor in the inactivation of wild-type p53 [20, 21]. Disruption of the interaction between wild-type p53 and MDM2/MDMX may release and reactivate p53 as a promising approach in cancer therapy (Fig. 3.3). Extensive research has been done on small molecular leads, such as nutlin-3a and RG7112 [22, 23], that

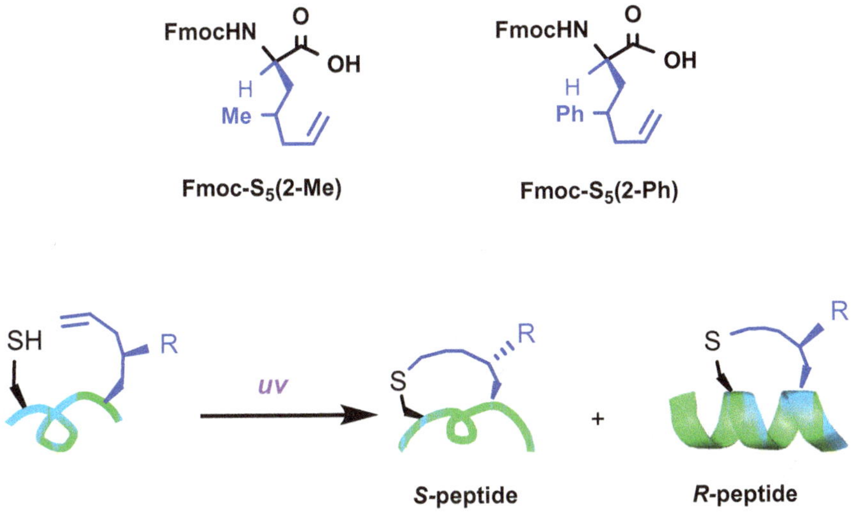

Fmoc-$S_5$(2-Me) Fmoc-$S_5$(2-Ph)

*S*-peptide *R*-peptide

**Fig. 3.2** A chirality-induced helicity system was used to construct stabilized peptides. The nonnatural amino acids used were synthesized following previously reported methodology

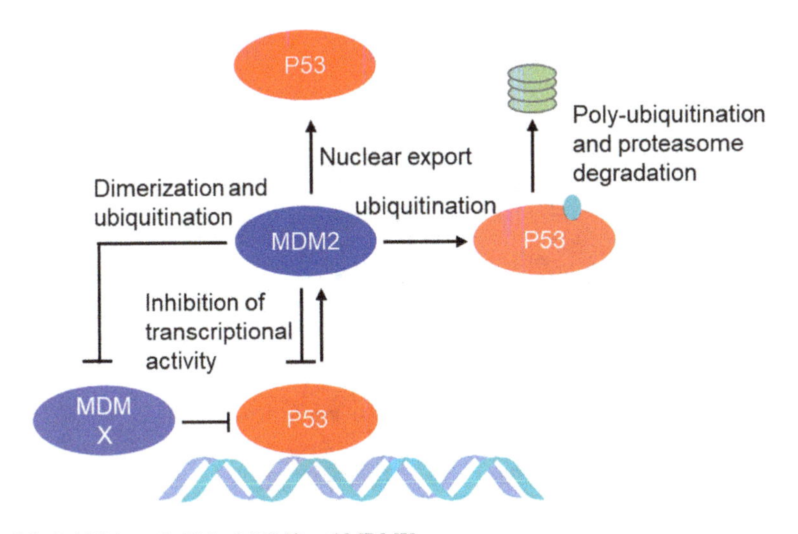


**Fig. 3.3** Inhibition of p53 by MDM2 and MDMX

target p53/MDM2 interactions. However, one potential limitation of these molecules is that they are all practically inactive against MDMX. Sawyer et al. reported a potent and selective small molecule, R-5963 that effectively inhibited p53 binding to both MDM2 and MDMX [24]. However, poor pharmacological characteristics render it unsuitable for further development. Recently, the use of stapled peptides has provided a new way to approach these typical PPIs targets, through converting the p53 helix from the native p53-MDM2/MDMX complex to a suitably stable, potent, and specific therapeutic agent. Verdine et al. reported a helical stabilized SAH-P53-8 peptide which reactivated p53 in vitro with limited biological potency in cellular assays [15]. Lane et al. reported a phage-derived sMTide02-02A peptide which displayed greater activity than SAH-8 and nutlin-3a [25]. Sawyer et al. reported on the stapled peptide ATSP-7041, a highly potent dual inhibitor of both MDM2 and MDMX with both in vitro and in vivo activity [26]. Although great successes in the modulation of p53-MDM2/X interaction by peptide-based ligands have been achieved, only limited in vivo studies were reported and more research is required to tackle this important PPI.

The resistance of cancer stem/progenitor cells (CSPCs) to chemotherapy leads to cancer relapse. Ovarian teratocarcinoma (OVTC) arises from germ cells and comprises pluripotent cells [27–29]. The PA-1 cell line, derived from human ovarian teratocarcinoma cells, is a well-accepted model for studying cancer cell stemness and expresses endogenous, nonfunctional wild-type p53 [30–32]. In this section, we designed CIH peptide modulators that target p53-MDM2/MDMX PPIs in CSCs with suitable cell permeability and binding affinity to reactivate the p53 apoptosis pathway and eradicate malignant CSCs (Fig. 3.4). In assessing the bioactivity of peptide drug leads, we thought that conducting in vivo animal model experiments in addition to

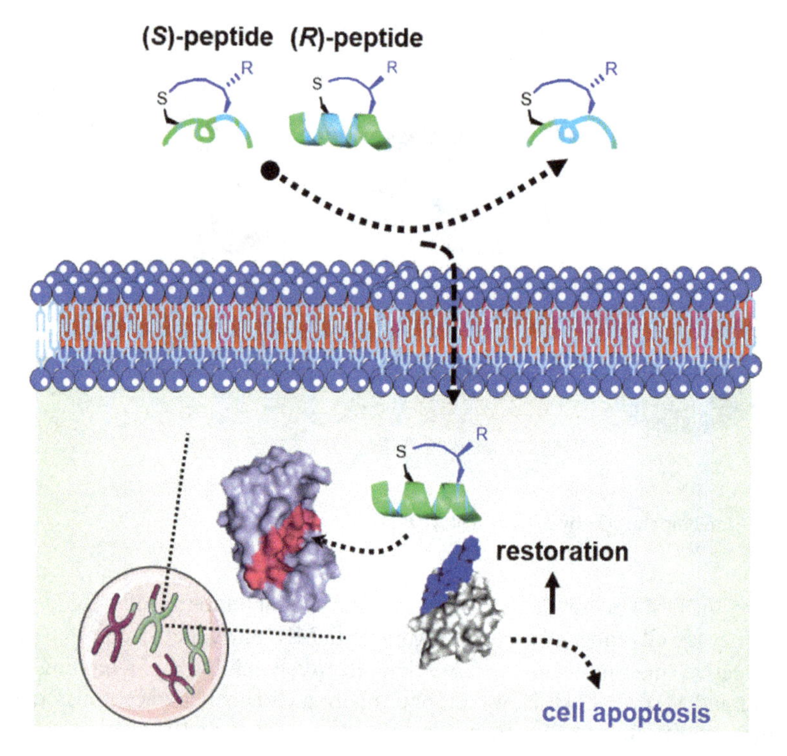


**Fig. 3.4** Designed CIH peptides inhibit the binding of MDM2/X to P53

in vitro cell-based experiments would provide more convincing results than the latter alone. In this study, for the first time, we assessed the peptide drug lead efficiency both in vitro and in vivo in CSCs.

## 3.2 Results

### 3.2.1 *MeR/PhR Is a Dual Inhibitor of MDM2 and MDMX and Exhibits Cell Nucleus Accumulation*

A p53 mimetic-stabilized peptide library was designed based on a recently reported dual inhibitor peptide (pDI) of MDM2/MDMX with the initial sequence of LTFEHY-WAQLTS discovered via phage display which named as PDI-1 [33]. Previous structural elucidation of MDM2 and its stapled peptide ligands suggested that the peptide tether also interacts with the hydrophobic cleft of MDM2/X [26, 34]. We performed iterative optimization and synthesized a library of stapled i, i + 4 peptides (entry 2–12) with the tether at different positions, different substitution groups (Methyl- or Phenyl-) at the chiral center, and amino acid mutations at selected positions (Table 3.1). For each peptide, it has two diastereomers—*S* and *R*, for which are different at the side chain chiral center. All peptides retained the core binding triad of Phe [19], Trp [23], and Leu [26] referring to the crystal complex of p53-MDM2 [29, 35]. However, unique contributions also exist, including (i) the staple moiety of peptides, which binds to the hydrophobic cleft of MDM2, (ii) the Tyr of PDI, which has been identified as a fourth key amino acid by several phage display studies [36, 37].

Fluorescence polarization assays were used to measure the binding affinity between the FITC-labeled peptide and GST-labeled MDM2/MDMX proteins (**Supplementary Fig. 1**). MeR and PhR exhibited the highest binding abilities toward both MDM2 and MDMX (for the entry number of peptides, unless noted otherwise, entry 2–12 refer to the *R* diastereomer peptides). The relatively low binding affinity may be caused by different MDM2/MDMX expression system, as both the CIH peptide MeR and PhR showed a greater binding affinity than the linear peptide PDI-1, which was reported as low as 10 nM in reference [38] but measured as 1 $\mu$M with proteins expressed in our lab. Notably, all the *R* diastereomers have a higher binding affinity than their respective S diastereomers, which unambiguously showed the importance of maintaining peptides' secondary structures (Fig. 3.5).

To investigate the cellular uptake of MeR/PhR, FITC-labeled TAT (FITC—GRKKRRQRRRPQ), a commonly used cell-penetrating peptide derived from HIV integrase[39], was used as a cellular uptake control for MeR and PhR. PA-1 cells were treated with 5 $\mu$M peptides and imaged at 2 hours post-treatment with confocal microscopy. Fluorescence-activated Cell Sorting (FACS) was used to quantify the fluorescent intensity in PA-1 cells. The results were summarized in Fig. 3.6. MeR and PhR showed diffused intracellular localization, while TAT-treated cells showed

**Table 3.1** Entry 2–11 were designed with different tether positions, tether type, and mutations at indicated positions. $K_d$ was measured with fluorescence polarization assays using FITC-labeled peptide R diastereomers with MDM2 and MDMX

| Entry | Sequence | | | | | | | | | | | | | | | $K_d$ ($\mu$M) | |
|---|---|---|---|---|---|---|---|---|---|---|---|---|---|---|---|---|---|
| | | | | | | | | | | | | | | | | $MDM_2$ | $MDM_X$ |
| 1 | L | T | F | Q | H | Y | W | A | Q | L | T | S | | | $NH_2$ | 1.40 | ND |
| 2*(MeR) | L | T | F | C | H | Y | W | $S_5$(Me) | Q | L | T | S | | | $NH_2$ | 0.16 | 0.53 |
| 3*(PhR) | L | T | F | C | H | Y | W | $S_5$(Ph) | Q | L | T | S | | | $NH_2$ | 0.50 | 0.66 |
| 4R | L | T | F | Q | C | Y | W | A | $S_5$(Me) | L | T | S | | | $NH_2$ | 0.56 | ND |
| 5R | L | T | F | Q | C | Y | W | A | $S_5$(Ph) | L | T | S | | | $NH_2$ | 0.61 | |
| 6R | L | C | F | Q | H | $S_5$(Me) | W | A | Q | L | T | S | | | $NH_2$ | >1.60 | |
| 7R | | | F | C | H | Y | W | $S_5$(Me) | Q | L | T | S | A | A | $NH_2$ | >1.31 | |
| 8R | L | T | F | C | H | Y | W | $S_5$(Me) | Q | L | T | S | A | A | $NH_2$ | 1.06 | |
| 9R | L | T | F | C | H | Y | W | $S_5$(Ph) | Q | L | T | S | A | A | $NH_2$ | >1.56 | |
| 10R | L | T | F | Q | C | Y | W | A | $S_5$(Me) | L | T | S | A | A | $NH_2$ | >1.50 | |
| 11R | L | T | F | Q | C | Y | W | A | $S_5$(Ph) | L | T | S | A | A | $NH_2$ | >4.14 | |
| 12R | L | S | F | C | Q | Y | W | $S_5$(Me) | Cba | L | S | P | | | $NH_2$ | >1.36 | |

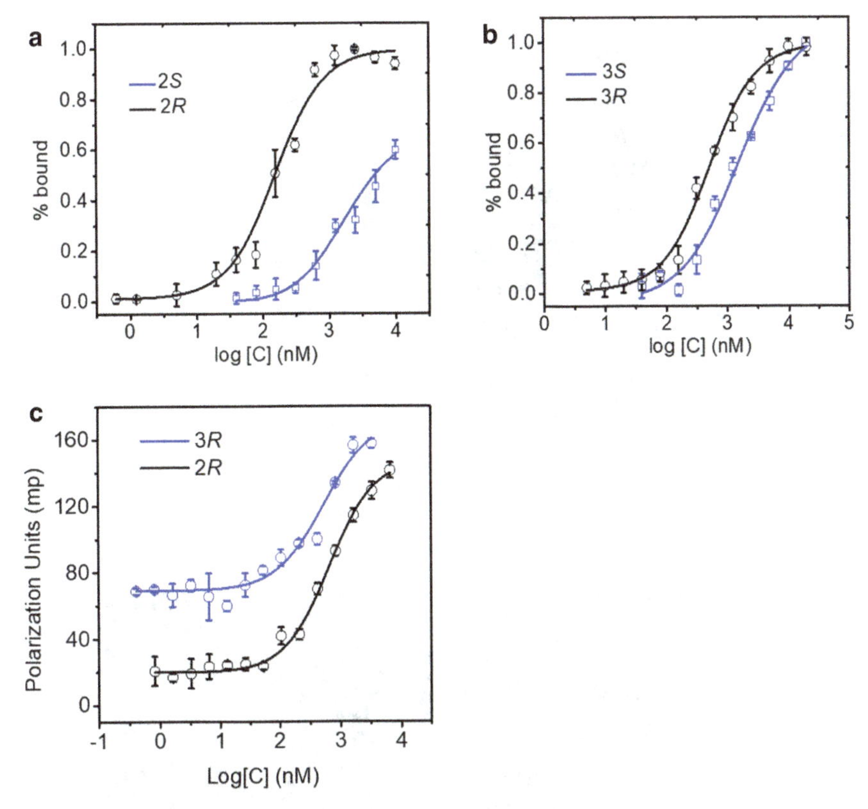


**Fig. 3.5** Fluorescence polarization assay to determine the $K_d$ of fluorescein-labeled peptides 2*S*/*R* and 3*S*/*R* bind to protein MDM2 and 2*R*/3*R* bind to protein MDMX. Data were demonstrated the *S* diastereomers have higher $K_d$ value compared to *R* diastereomers of each peptide. MeR(2R) and PhR(3R) were screened to show the best affinity to MDM2/X

almost no detectable intracellular fluorescence. PhR showed higher cellular uptake than MeR in both PA-1 cells.

Immunofluorescence was used to study the co-localization of the peptides and MDM2. The MDM2 protein was stained with an Alexa-647-labeled antibody and was mainly located in the nucleus. The results showed that PhR can accumulate in the nucleus and bind to MDM2/X (Fig. 3.7).

### 3.2.2 MeR/PhR Activates P53 Signaling in Cancer Cells

Inhibition of p53 binding to its negative regulators, MDM2 and MDMX, could possibly stabilize p53 and activate the pathway. Activation of the p53 pathway only occurs in cells that express wild-type p53 but not in cells that express the mutant form of the p53. To assess the activity of MeR/PhR, we chose cancer cells lines representing

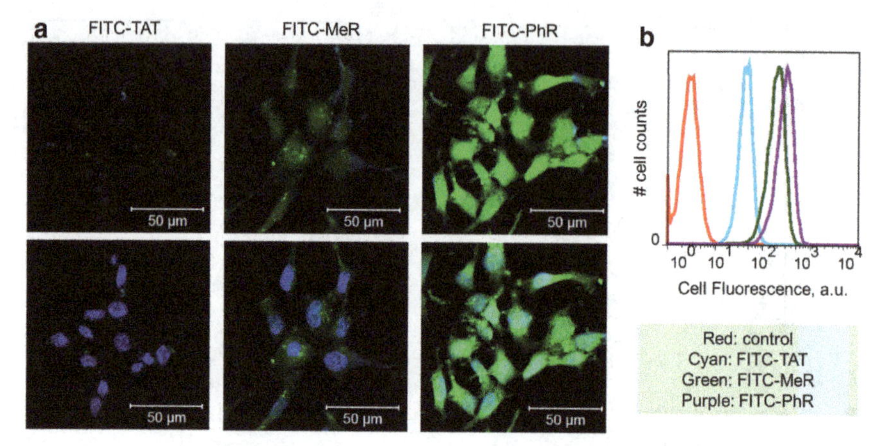


**Fig. 3.6** (**a**) Confocal microscopy images of PA-1 cells treated with 5 μM FITC-labeled peptides at 37°C for 2 h. Scale bar, 20 μm. DAPI: blue; peptide: green. (**b**) Flow cytometry analysis of PA-1 cells treated with 5 μM FITC-labeled peptides at 37°C for 2 h

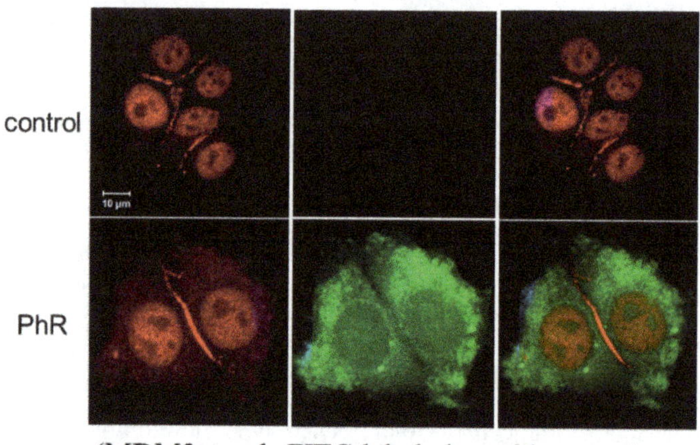


**Fig. 3.7** Immunofluorescence imaging demonstrated FITC-labeled peptides' uptake by MCF-7 cells and that they co-localized to the MDM2 protein in cell nucleus. MDM2 was stained with Alexa Fluor 647-labeled antibody(red)

two clinically relevant populations of tumors that overproduce either MDM2 (PA-1 ovarian teratocarcinoma) or MDMX (MCF-7 breast cancer) (Fig. 3.8a, b). First, we measured the $IC_{50}$ of our peptides using cell viability assays. Nutlin-3a was chosen as a positive control. Two mutant p53 cell lines (Skov-3 and MDA-231) were chosen to exclude the nonspecific toxicity of MeR and PhR. Negligible effects were detected for two cell lines' viability up to 50 $\mu$M of either MeR and PhR (Fig. 3.8c). A normal cell line (HEK-293) was also tested and negligible cytotoxic effects were detected with peptide treatment as shown in Fig. 3.8d. In contrast, treatment of MCF-7 and

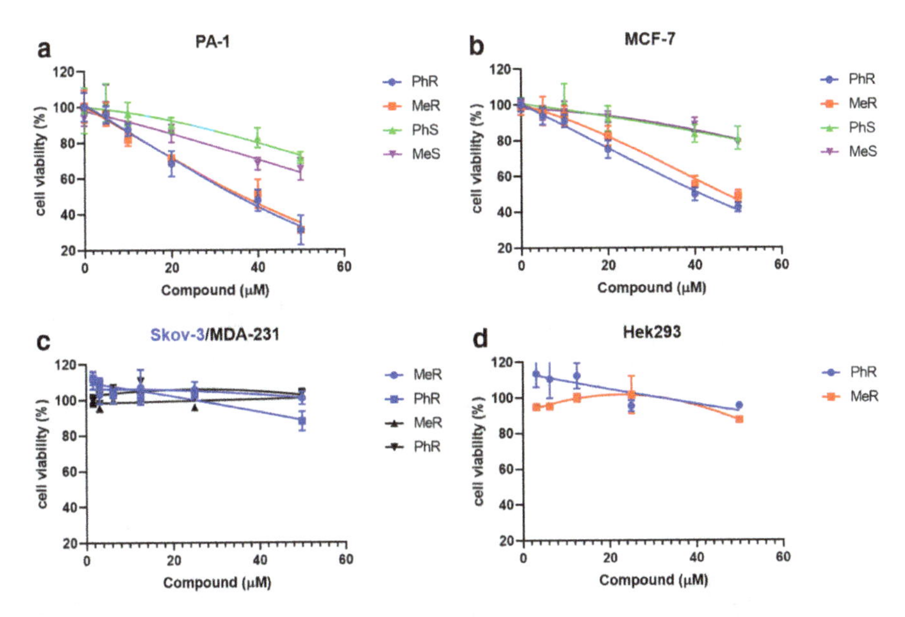


**Fig. 3.8** Viability of two p53wt cancer cells (MCF-7 and PA-1), two p53 mutant cancer cells (MDA-231 and skov3), and a normal cell line (Hek-293). PA-1 and MCF-7 cells were incubated with 5, 10, 20, 40, and 50 μM MeR, PhR, or their diastereomers-MeS and PhS for 48 hours. Skov-3, MDA-MB-231, HEK-293, and QSG-7701 were incubated with 3.12, 6.25, 12.5, 25, and 50 μM MeR or PhR for 48 hours. Both MeR and PhR had little effect to the cell growth and proliferation. Error bars represent SEMs for triplicates of the data

PA-1 cells with MeR and PhR led to significant cell growth inhibition in a strictly dose-dependent manner ($IC_{50} = 40\ \mu M$) (Fig. 3.8a, b). From the bright field images of cells treated with MeR/PhR, we observed apparent cell proliferation inhibition in a time- and dose-dependent manner (Fig. 3.9). These results proved the target specificity of the peptides MeR and PhR in the p53wt stem-like cancer cell line and normal cancer cell lines. To exclude the lack of cell death due to the lack of peptide internalization in P53-mutant cell lines, we performed cellular uptake assays and proved that the level of cellular uptake of PhR in MDA-231 is similar to that of PA-1 (Fig. 3.10).

The levels of protein and mRNA expression were measured to assess the biological activity of our peptides. PA-1 and MCF-7 cells were treated with MeR, PhR, and nutlin-3a for 48 h, and then p53, MDM2 and MDMX expression levels were monitored via western blot analysis (Fig. 3.11). Notably, p53 expression was upregulated with MeR and PhR treatment. The peptides' S diastereomers showed negligible effects.

The activation of p53 by peptide MeR and PhR was also demonstrated with the induction of the mRNA levels of three p53 target genes, MDM2, MDMX, and MIC-1 in both PA-1 and MCF-7 (Fig. 3.12). Direct evidence of CSC inhibition came from the downregulation of stemness genes and upregulation of differentiation genes in CSCs after treatment. Indeed, the reactivation of p53 via PhR treatment resulted in

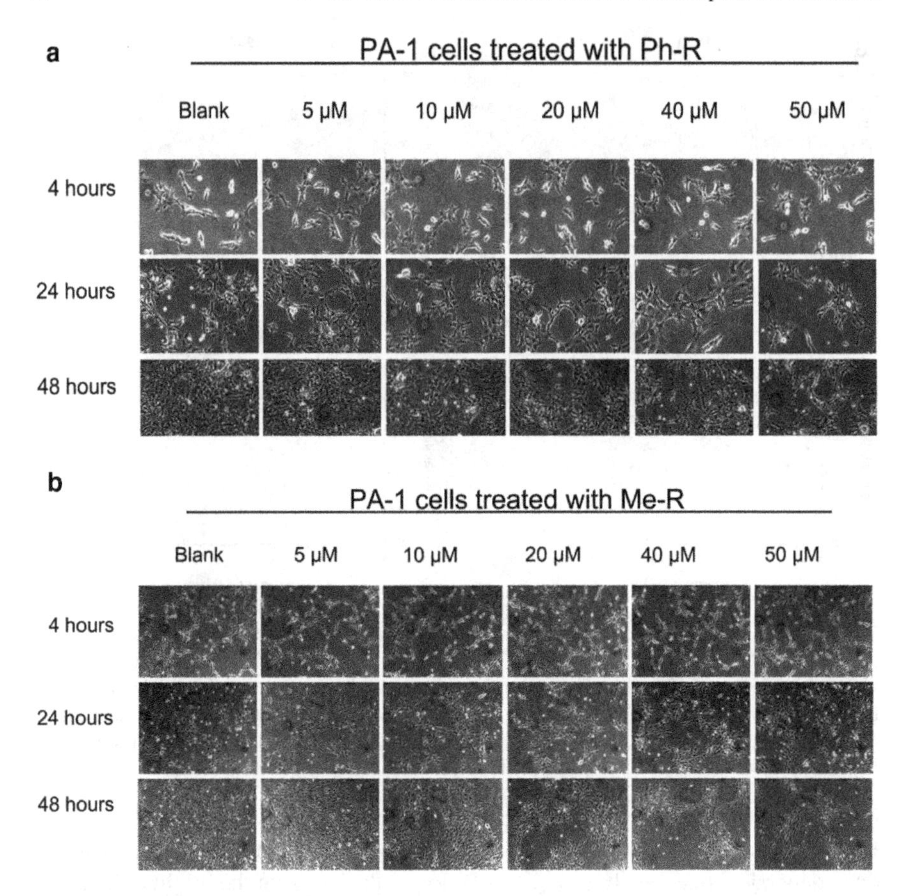


**Fig. 3.9** Bright field images of PA-1 cells treated with MeR/PhR showing a time and dose-responsive effect to cell death. Apparent cell death was observed in PA-1 cells when incubated with 50 μM MeR or PhR for 48 h

significant upregulation of the differentiation gene foxA2 and downregulation of the stemness gene sox2 (Fig. 3.12a), which further suggested that the deactivation of p53 could play an essential role in the proliferation of CSC cells. Similar changes were observed in the expression of sox2 and foxA2 in MCF7 cells [39] treated with PhR (Fig. 3.12b). In summary, PhR served as an efficient regulator that reactivated p53 to regulate pluripotency related genes. Notably, changes in gene regulation caused by peptide treatment were observed only in p53wt cancer cells (PA-1 and MCF-7) and not in normal (QSG-7701) cells (Fig. 3.12c).

The S diastereomers of MeR and PhR, MeS and PhS were used as control peptides, a practice that circumvented the use of scrambled or mutated peptides as negative controls. Their results were shown in Fig. 3.13, labeled as MeS and PhS. The functional difference between the peptide diastereomers clearly indicated the importance

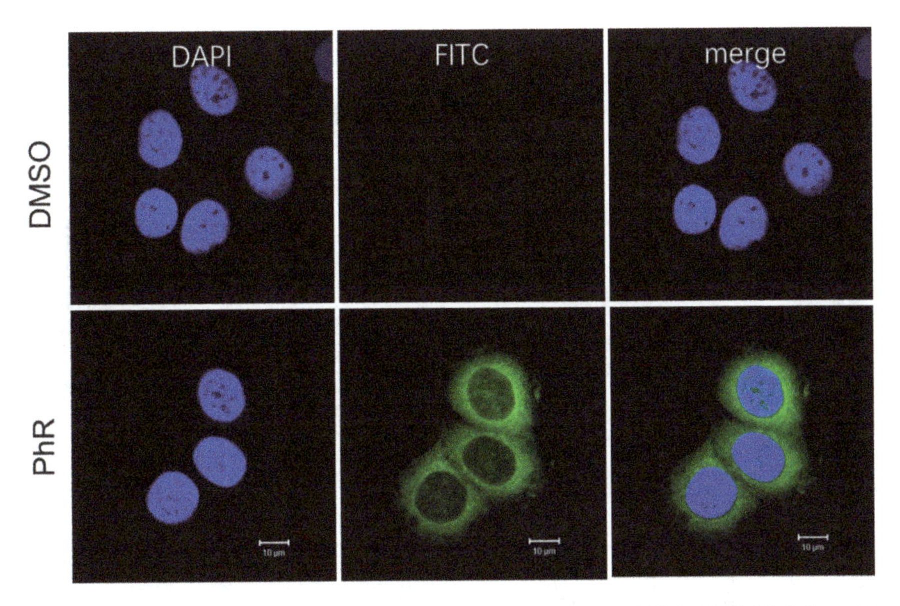


**Fig. 3.10** Confocal microscopy of MDA-231 cells treated with FITC-PhR. FITC-labeled peptide was incubated with MDA-231 cells for 2 h at 37°C. DAPI (blue), FITC (green)

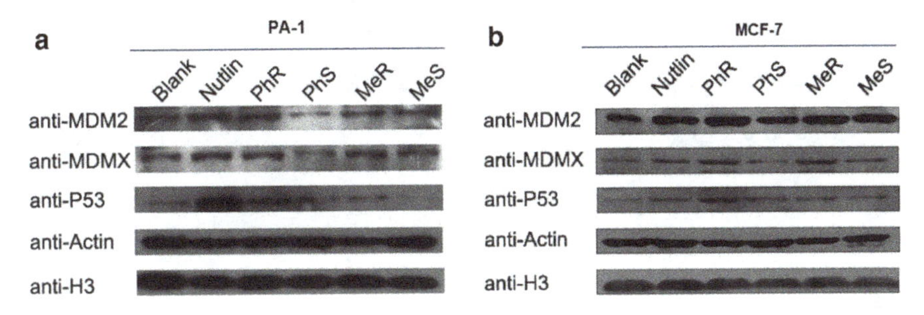


**Fig. 3.11** MeR/PhR stabilizes p53 and elevates protein levels of p53 and its targets MDM2 and MDMX. Log-phase MCF-7 and PA-1 cells were incubated with 40 μM peptides (MeR/S, PhR/S) or 1 μM nutlin-3a, and cell lysates were analyzed by western blotting

of tether structure. Our CIH strategy provides an ideal platform for investigating the correlations between peptide function and conformation.

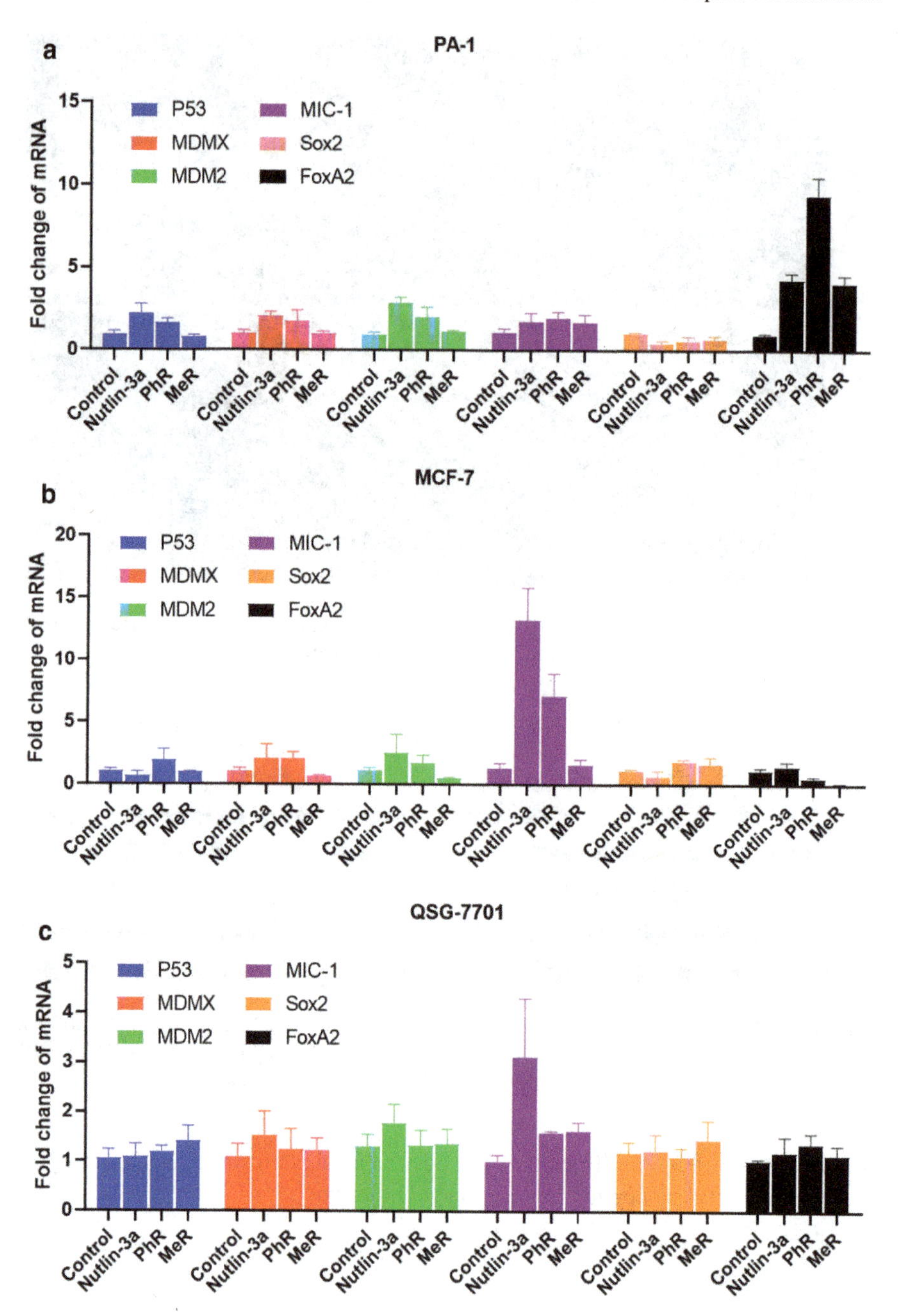


**Fig. 3.12** PhR shows induction of p53 and p53 target genes in p53wt cell lines. Exponentially growing p53wt cancer cell lines were incubated with 40 μM peptides or 1 μM nutlin-3a for 48 h, and mRNA level of p53 and p53 targets MDM2, MDMX, and MIC or stemness-related genes were analyzed by quantitative PCR and expressed as fold increase. Changes of gene expression in (**a**) PA-1, (**b**) MCF-7, and (**c**) QSG-7701. Error bars represent SEMs for triplicates of the data, **, P<0.01. *, P<0.05

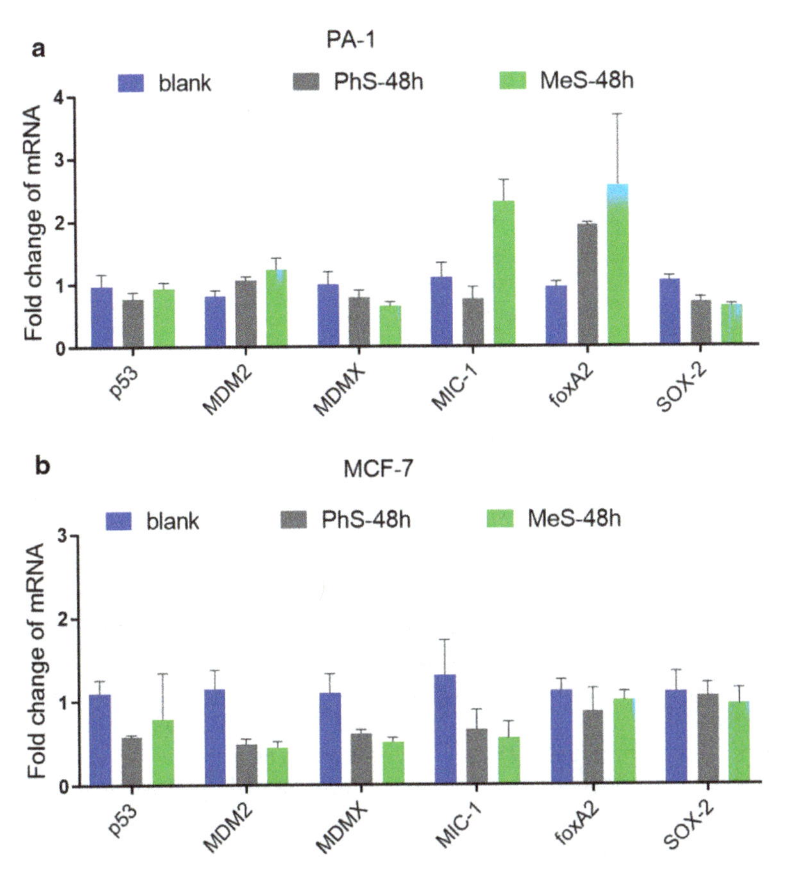


**Fig. 3.13** Exponentially growing p53wt cancer cell lines were incubated with peptides MeS/PhS(40 μM) for 48 hours, the p53 and p53-related genes expression levels were analyzed by quantitative PCR and expressed as fold increase. Changes of gene expression in (**a**) PA-1 and (**b**) MCF-7. Error bars represent SD for triplicates of the data

### *3.2.3 MeR/PhR Reactivates Major P53 Cellular Functions in Cancer Cells that Overexpress MDM2 and MDMX*

The tumor suppressor p53 possesses many cellular functions, the most important of which is the induction of apoptosis and cell-cycle arrest [36, 40, 41]. PA-1 and MCF-7 cells were exposed to 40 $\mu$M peptides or 5 $\mu$M nutlin-3a for 48 h, the latter of which was known to induce apoptosis by inhibiting MDM2/MDMX [35, 37, 42]. Annexin-V/PI assays were used to quantify the apoptotic effect of MeR and PhR. The apoptotic effect of PhR was more pronounced in both PA-1 and MCF-7 cells than those of MeR and nutlin-3a (Fig. 3.14); this was consistent with the viability results.

MCF-7 cells are known to exhibit caspase-3 deficiency and a partially compromised apoptotic response, and the lower Annexin V levels are consistent with this

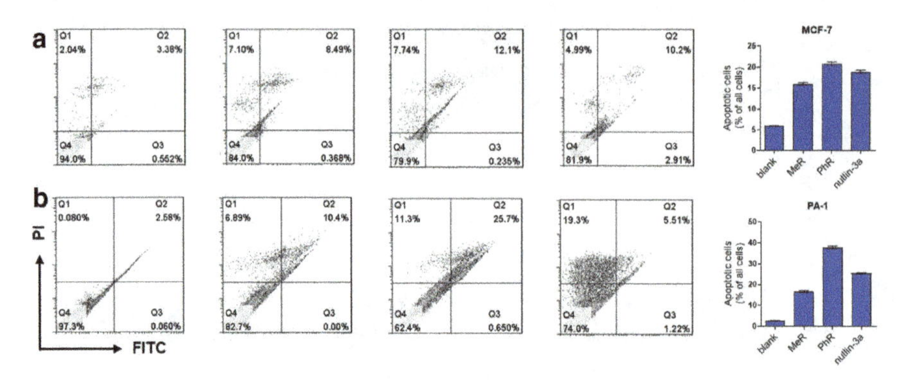


**Fig. 3.14** MCF-7 and PA-1 cancer cell lines were treated with MeR, PhR (40 $\mu$M) and nutlin-3a (5 $\mu$M) for 48 h. The percentage of apoptotic cells was determined through the PI/Annexin V double stain assay, the red cycles in both **a** and **b** showed PhR induced higher apoptosis in both cells than MeR or nutlin-3a. The red rectangle in **b** showed higher percent of cell death when treated with nutlin-3a. The column graph was used to show the ratio of apoptotic cells in total cells. The apoptotic cells were calculated by Q1 + Q2 + Q3

understanding. It is worth noting that nutlin-3a led to more cell deaths through means other than apoptosis, which may have been caused by the nonspecific cytotoxicity of nutlin-3a. Notably, PhR was more effective in inducing cell-cycle arrest in the G2/M phase than MeR and nutlin-3a in PA-1 cells (Fig. 3.15a). In MCF-7 cells, nutlin-3a caused profound G0/G1 cell-cycle arrest as previous literature reported [43, 44], whereas PhR exhibited diverse effects on cell-cycle arrest (Fig. 3.15a), which suggests that PhR activated the p53 pathway via a heterogeneous mechanism. It is worth noting that PhR did not induce cell-cycle arrest in the normal cell line

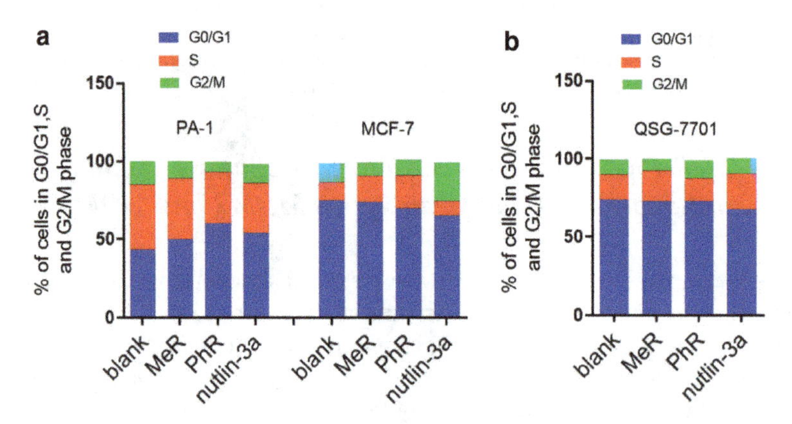


**Fig. 3.15** Cell-cycle distribution was determined through PI staining in PA-1 cells after treated with MeR, PhR (40 μM), or nutlin-3a (5 μM) for 48 hours. The flow cytometry data were analyzed with FlowJo software. Column graph to show the distribution of cell-cycle phase for (A) PA-1 and MCF-7 and (C) QSG-7701 after treated with MeR, PhR, or nutlin-3a

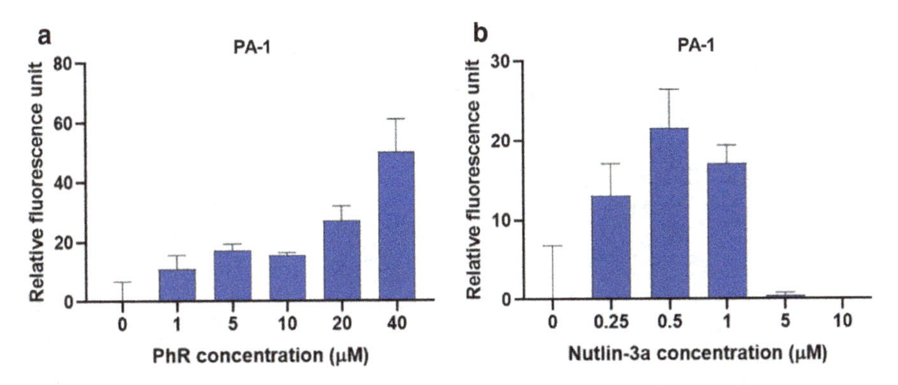


**Fig. 3.16** (**a**) PhR demonstrated dose-dependent apoptosis induction as measured by a caspase-3 assay. The extent of apoptosis was measured through the detection of caspase-3 activity by exposing the cells to a caspase-3-specific substrate. (**b**) caspase-3 activity in PA-1 cells treated with nutlin-3a

QSG-7701 (Fig. 3.15b), suggesting that PhR is more specific and less toxic than nutlin-3a.

The caspase-3 assay indicated that PhR induced dose-dependent apoptosis in PA-1 cells (Fig. 3.16a). Nutlin-3a only induced minimal apoptosis in a low dose, while in higher doses (>5 $\mu$M), no caspase-3 activity was detected (Fig. 3.16b). These results further indicated nutlin-3a killed cancer cells partly through its nonspecific toxicity rather than all through p53 pathway activation.

MDM2 negatively regulates p53 function through multiple mechanisms, including direct binding, which masks the p53 transactivation domain, impairing nuclear import of the p53 protein, and ubiquitination and proteasomal degradation of the p53 protein. Reactivating p53 by inhibiting MDM2 inhibits the ubiquitination of p53 and blocks its export from the nucleus for degradation [45]. GFP-labeled p53 and MDM2 plasmids were co-transfected into HCT-116 cells for 24 h; the cells were then treated with peptides for 24 h. The confocal image results showed that p53 ubiquitination was inhibited when treated with PhR, as no green fluorescence was observed in the cell plasma (Fig. 3.17), while PhS had little effect to this process. Here we further confirmed our peptides' capacity for nuclear penetration.

### *3.2.4 PhR Peptide Upregulates P53 and Induces Cell Apoptosis in a Time-Dependent Manner*

PhR activated the p53 pathway and induced cell apoptosis. PA-1 cells were treated with a dose ladder from 0 $\mu$M to 40 $\mu$M with 10% serum. After 2 days, a dose-dependent increase of the level of p53 and the concomitant elevation of the p53 transcriptional targets MDM2 and MDMX were observed. The activation of p53 by PhR was demonstrated by the dose-dependent induction of the mRNA of p53 and three p53 target genes—p21, MDM2, and MDMX (Fig. 3.18a). The response time of

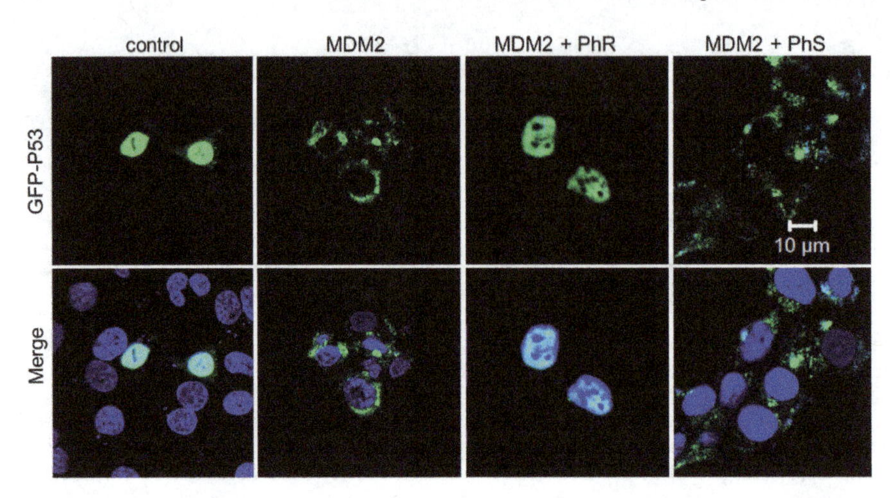


**Fig. 3.17** PhR inhibits the ubiquitination of p53 by MDM2. GFP-labeled p53 and MDM2 plasmids were co-transfected into HCT-116 cells for 24 h, which were then treated with peptides for 24 h. Images were taken by a Zeiss confocal microscopy. In control, p53 major locates in cell nucleus, with co-transfection with MDM2, the p53 was translocated to cell plasma. PhR effectively inhibited the translocation process in 24 h. GFP-p53: green; DAPI: blue

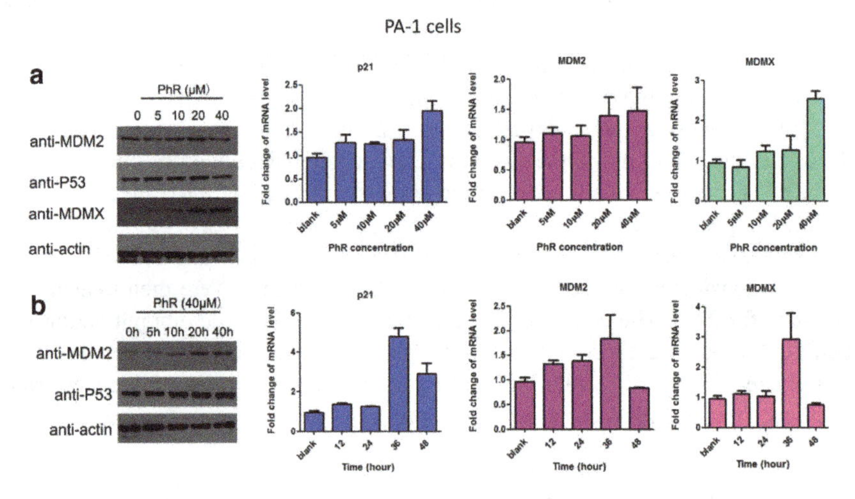


**Fig. 3.18** PhR reactivates the p53 pathway in a time- and dose-dependent manner. (A) PhR elevates p53 and p53 targets MDM2/X protein level and shows induction of the p53 target genes p21 and MDM2 in a dose-dependent manner in PA-1 cells. Log-phase PA-1 cells were exposed to 5, 10, 20, and 40 $\mu$M PhR for 48 h and cell lysates were analyzed by western blotting. Exponentially growing p53wt cancer cell line PA-1 was incubated with 5, 10, 20, 40 $\mu$M PhR for 48 h for quantitative PCR. (B) PhR elevates p53 protein and p21 protein levels in a time-dependent manner. The mRNA levels of p21 and MDM2 changed as the incubation time changed. For western blotting, log-phase PA-1 cells were incubated with 40 $\mu$M PhR for 5 h, 10 h, 20 h, and 40 h and cell lysates were analyzed by western blotting. For quantitative PCR, PA-1 cells were incubated with PhR for 12, 24, 36, and 48 h. (C) Transfected HCT-116 cells treated with nutlin-3a(0.5 $\mu$M) and PhR(40 $\mu$M), changes in the protein expression and mRNA levels in a time-dependent manner were detected

p53 activation after treated with PhR was tested by using a fixed peptide concentration of 40 $\mu$M from 0 to 48 hours. Apparent p53 and MDM2 accumulation both in protein and mRNA levels was detected, with the highest mRNA level observed slightly before the highest protein level (Fig. 3.18b). Recently, Lahav et al. reported that cell death depends on the accumulation kinetics of p53 [46]. With HCT-116 Venus knock-in cell line (p53-VKI) (a kind gift from the Lahav laboratory), we detected a more rapid accumulation of p53 protein in PhR-treated cells than in nutlin-3a treated cells. For PhR, the highest levels of p53 and p21 were reached in 12 hours, while that for nutlin-3a was 48 hours (Fig. 3.19a), which clearly indicated a delay as compared with PhR. This phenomenon may help to explain PhR's working mechanism and its greater efficacy in vivo compared with nutlin-3a. The faster accumulation of p53 in PhR-treated cells was further confirmed via measurements of the mRNA levels of p53 and related transcriptional genes (Fig. 3.19b).

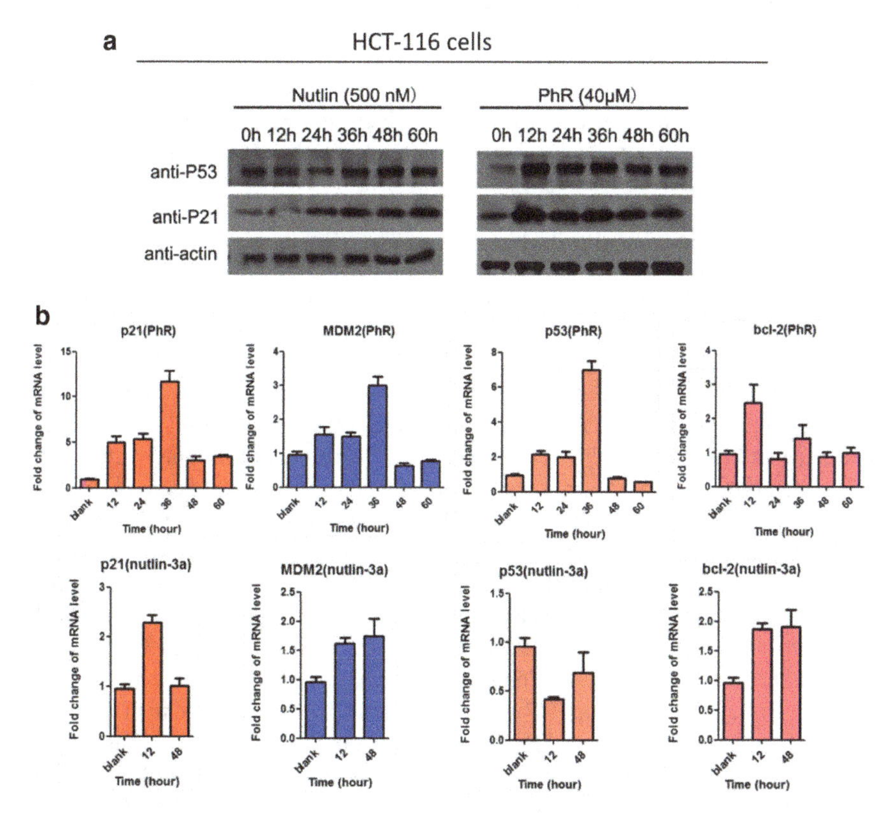


Fig. 3.19 The P53 and p53-related genes expression level in HCT-116 (p53-VKI) cells at different time points when treated with PhR(40 $\mu$M) or nutlin-3a(5 $\mu$M). (a) p53 (b) caspase-3 (c) MDMX (d) bcl-2. HCT-116 Venus knock-in cell line (p53-VKI) was a kind gift from the Lahav laboratory

### *3.2.5 Transcriptome Analysis of PhR-Treated Cells*

To obtain an overview of the genes directly affected by activation of p53 pathway, genome-wide mRNA microarrays were performed for PA-1 cells treated with PhR. Compared with untreated cells, a total of 285 downregulated and 93 upregulated genes were detected (fold change > 2) and clustered in PA-1 cells exhibiting the activation of p53 pathway (Fig. 3.20a–d). Despite this result, gene ontology analysis revealed the number of genes with changed expression levels is involved in different pathways, such as in the signaling pathway regulating the pluripotency of stem cells, the p53 signaling pathway, the Ras signaling pathway, the PI3k-Akt signaling pathway, and the Rap signaling pathway, and so on (Fig. 3.20e). These genes play important roles in regulating the activity of cancer cells and the results are summarized in Fig. 3.20f. Together, these results demonstrate that PhR peptides act as potent modulators of p53-MDM2/X pathway and activate cell apoptosis.

### *3.2.6 PhR Suppresses Tumor Growth in Vivo in PA-1 Xenograft Model via Reactivation of the P53 Pathway*

PhR was examined for its efficacy in a human cancer xenograft model carrying p53wt and elevated levels of MDM2 proteins, namely the PA-1 human xenograft model. As shown in Fig. 3.21, PhR resulted in statistically significant tumor growth inhibition (TGI) in the MDM2-amplified human teratocarcinoma xenograft model PA-1. When dosed for 3 weeks at 10 mg/Kg every other day with intratumoral injections, PhR induced a TGI of 70% (Fig. 3.21a). In comparison, the selective MDM2 small-molecule inhibitor nutlin-3a (administrated every other day at doses of 10 mg/Kg) resulted in a TGI of 30%, which is much less effective compared to PhR. Actually, the drug concentration in the tumor section is much higher for nutlin-3a than PhR for this dosage [M.W.: PhR (1640) VS nutlin-3a (581)]. These results are confirmed by tumor weight and tumor volume measurements (Fig. 3.21b, c). At the same time, the weight of the mice showed no significant change (Fig. 3.21d), suggesting that our peptides have little side effect with respect to mouse growth. Furthermore, the mortality of the mice during the survival study was shown as survival curves (Fig. 3.21e), which indicate that PhR was able to prolong the lives of the mice. Especially, no mice died during the treatment period in PhR group while three in ten mice died in the nutlin-3a group. To assess the pharmacokinetics and duration time of PhR within the tumor, Cy3-labeled PhR was treated by intratumoral injections and imaged at different time points (0 hour, 5 min, 30 min, 2 hours, 6 hours, and 24 hours) using the IVIS Lumina II small animal in vivo optical imaging system. The optical signal intensity of the PhR-Cy3 peptide remained consistently without apparent diminishing for 24 hours (Fig. 3.21f and g). Furthermore, the ex vivo fluorescence results from tumors and major organs were consistent with previous reports on the biodistribution of peptides

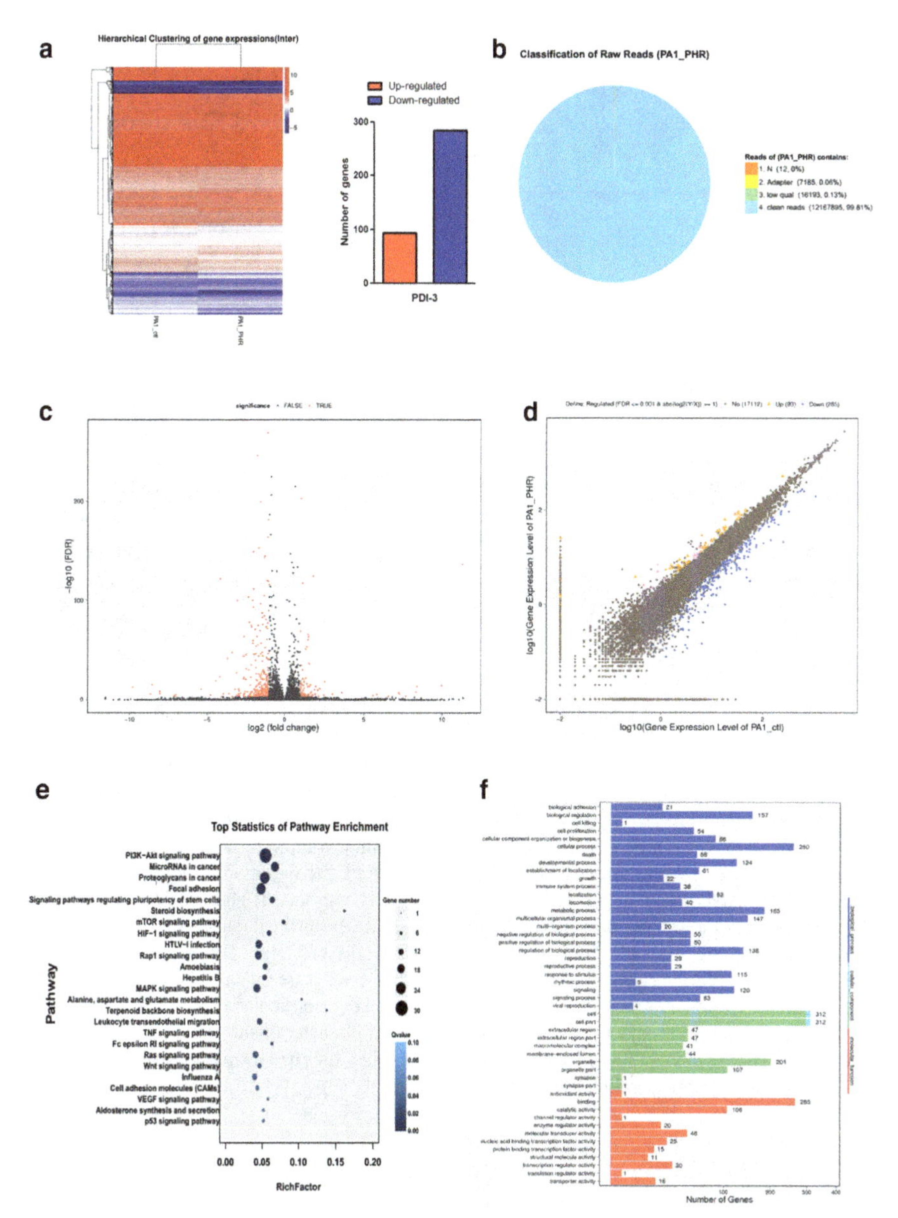


**Fig. 3.20** Transcriptome analysis of PA-1 cells treated with PhR and PBS as blank. (**a**) Heat-map of genes differentially expressed in RNA-microarray analysis performed on PA-1 cells treated with PhR compared with blank. Microarray analysis reveals a total of 285 downregulated and 93 upregulated genes in PA-1 cells (n = 3; fold change >2.0; $p$<0.05). (**b**) The quality of data. (**c** and **d**) 2D diagram to show the number of genes with change of expression level and gene ontology analysis in PA-1 cells treated with PBS as blank or PhR peptides. Gene expression level and higher ranked 20 changed pathways were listed for PA-1 cells treated with PhR peptides compared to PBS as blank. (**e**) Gene ontology analysis reveals the different pathways in which the changed genes are involved. P53 and several p53-related pathways are observed. (**f**) Gene ontology analysis revealed the different pathways for involving the changed genes

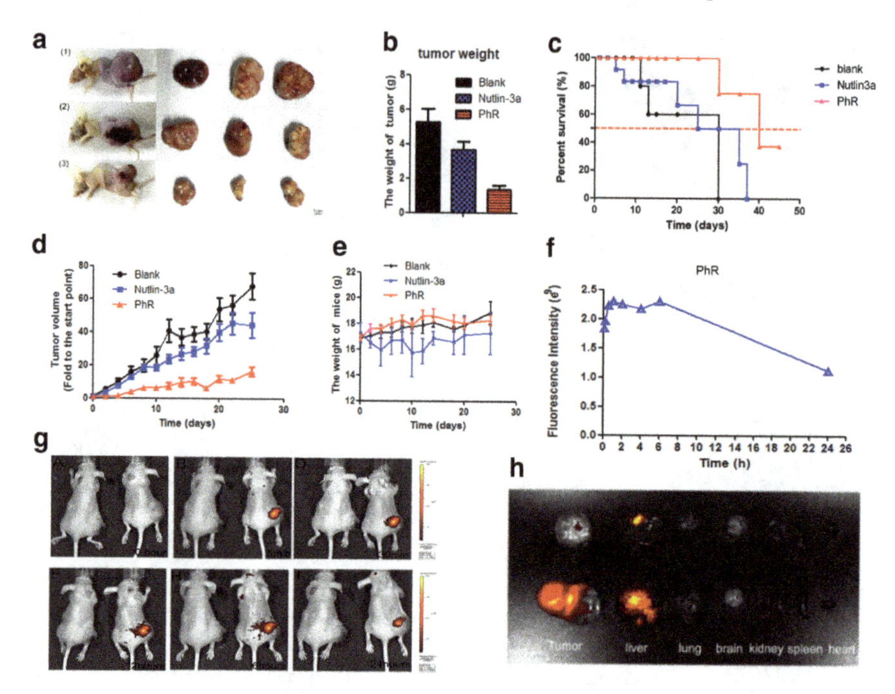


**Fig. 3.21** Antitumor activities of PhR in a PA-1 stem-like cancer cell xenograft animal model. (**a**) Tumor tissue image of mice treated with (1) PBS, (2) nutlin-3a, (3) PhR. Mice treated with PhR in the last group exhibited the average smallest tumors. (**b**) The weight of the tumors. Data were plotted as mean ± SEM. (**c**) Relative changes in tumor volume versus time. Tumor volumes were calipered throughout the study, and data were plotted as mean ± SEM. Relative tumor volume is defined as $(V-V_0)/V_0$, where V and $V_0$ indicate the tumor volume on a particular day and day 0, respectively. Error bars represent SEMs for triplicate data. Mean tumor volumes were analyzed using one-way ANOVA. Values are means ± SEM, n = 4−6 tumors. (**d**) The weight of the mice was measured during the treated time schedule. (**e**) Kaplan-Meier survival curves for epidermal graft tumor nude mice treated with PhR and Nutlin-3a for 21 days and raised for another 24 days post-treatment. The nutlin-3a caused 20% mice death in the first-week treatment. The PhR group has no mice death until stop treatment for one week. (**f**) The body distribution of PhR $^{Cy3}$ at different time points after intratumoral injections. (**g**) The optical signal intensity quantified from (f) at different time points. Half percent of intensity retained after 24 h injection. (**h**) The major organs' distribution of PhR-cy$^3$ after intratumoral injection for 24 hours

and indicate that the majority of the peptide accumulated in the tumor and liver/kidney (Fig. 3.21h).

Immunohistochemistry assay was used to check the p53 and target protein expression level. Notably, higher expression levels of p53 and p21 in PhR-treated tumor tissues were detected via immunohistochemistry (Fig. 3.22a, b), this is consistent with in vitro results. Caspase-3 and proliferating cell nuclear antigen (PCNA) detection via immunohistochemistry was used to determine the apoptosis level (Fig. 3.22c, d). Although nutlin-3a displayed a lower $IC_{50}$ than PhR in vitro, when the mice were

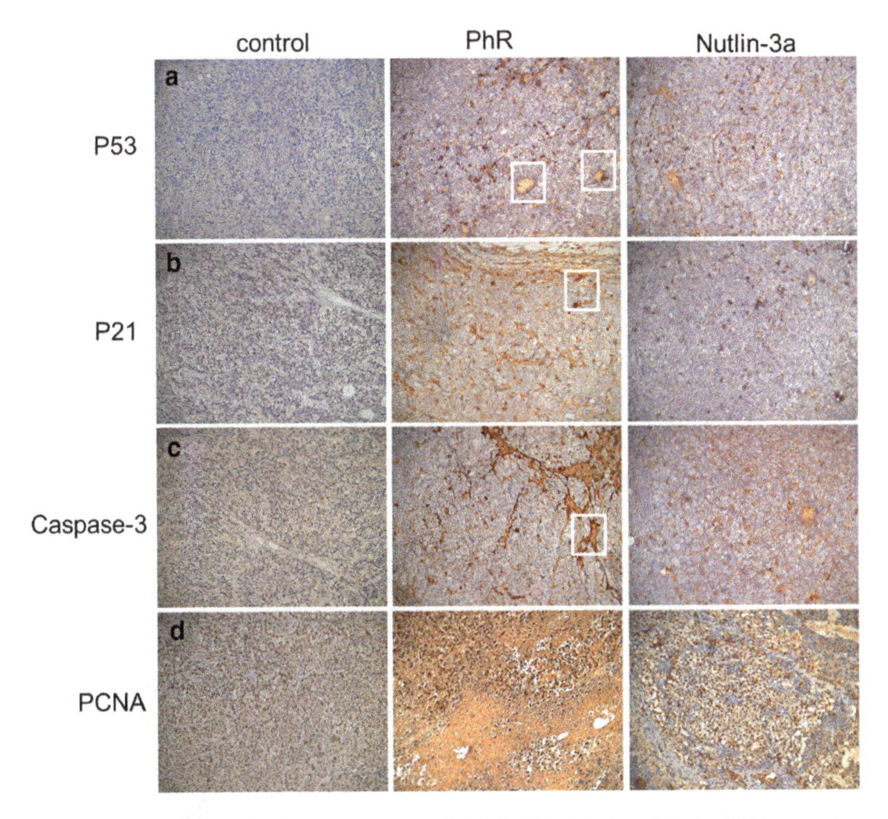


**Fig. 3.22 Immunohistochemistry analysis of p53 and p53-related protein level in tumor tissue slides.** Teratocarcinoma tissues collected from different groups of mice after 3 weeks treatment. (**a**) p53 (**b**) p21 (**c**) caspase-3 (**d**) PCNA were apparently elevated in the PhR-treated mice. The white rectangle in the slides indicates respective proteins. MicroSpot Focusing Objective, 20×

treated with higher doses (10 mg/Kg) in vivo, it induced a lower level of apoptosis and displayed less tumor inhibition than peptide (10 mg/Kg).

### *3.2.7 PhR Shows High Biocompatibility and Low Toxicity in Vivo*

According to the in vitro and in vivo experiments, PhR efficiently inhibited cancer growth in p53 function-deficient mice. However, low-toxicity and harmless elimination from the body within a reasonable timeframe are as important to consider as efficacy. A mice voluntary cage-wheel exercise assay was designed to assess the toxicity, and histological studies were performed for the biodistribution of PhR in vivo. In the mice voluntary cage-wheel exercise assay, BALB/c mice were randomly divided into two groups and were subcutaneously injected with PBS or PhR (10 mg/kg). Over a period of 20 days after injection, the voluntary running cycles increased steadily and no significant differences were found between these two groups, which indicates no obvious harmful effect to the motor learning ability of mice treated with PhR peptide (Fig. 3.23). Histological analysis was performed to further evaluate the toxicity of PhR in vivo. Major organs including the tumor, heart, liver, spleen, lung, kidney, and brain tissue sections were harvested for hematoxylin and eosin (H&E) staining to observe histological changes at 21 days after injection. The PhR group had a more apparently tumor inhibition effect compared to other groups (Fig. 3.24). Compared with the PBS-treated mice, no signs of organ lesions were observed in the mice treated with PhR. Therefore, these in vivo toxicity assessments strongly indicate PhR as a low-toxic and biocompatible modulator for cancer therapy.

## 3.3 Discussion and Conclusion

During the past decade, multiple strategies have been conceived with the specific aim of destroying CSCs and their niche [47]. These include targeting specific surface markers [48, 49], modulating signaling pathways [50], adjusting microenvironment signals [51], inhibiting drug-efflux pumps [52], manipulating miRNA expression [53], and inducing CSCs apoptosis and differentiation [54]. While these alternative therapies are promising, some of them may not be specific and affect healthy tissue, since the CSC niche is similar and located close to the normal stem cell niche, which can also be affected by the compounds [55, 56]. There are also many ways by which CSCs could evade treatment. The challenges to tackling CSCs include the improving specificity and efficiency in targeting CSCs, avoiding toxicity to normal tissue stem cells, and ensuring drug delivery and retention [48, 49].

Recently, we reported a CIH strategy that displays specific tunable characteristics in the chiral center and thus allows us to switch the biophysical properties of the peptides. We hypothesized that reactivating the p53 pathway in PA-1 cells, a well-accepted model for studying cancer cell stemness, with peptides designed based on the CIH concept would decrease their stemness and kill the cells. As a proof of concept, we showed that inhibiting the interaction between p53 and MDM2/X can inhibit the growth of stem-like cancer cells both in vitro and in vivo.

In summary, a potent dual peptide inhibitor based on the CIH concept was designed to specifically target MDM2/X and could further induce p53-dependent apoptosis and inhibit cell proliferation. Notably, the CIH strategy provides an ideal way to obtain peptides with identical chemical compositions as controls instead of mutated or scrambled control peptides used in all previous studies. The S PDI diastereomers were used as controls for target binding affinity, cellular uptake, cell viability, cell-cycle arrest, apoptosis induction, and protein/mRNA regulation. The results clearly emphasize the decisive correlation between the peptides' secondary structure and their functions.

Although PhR showed comparatively weaker cellular activity than nutlin-3a in cell assays, it exhibited superior in vivo efficacy within the pluripotency cancer

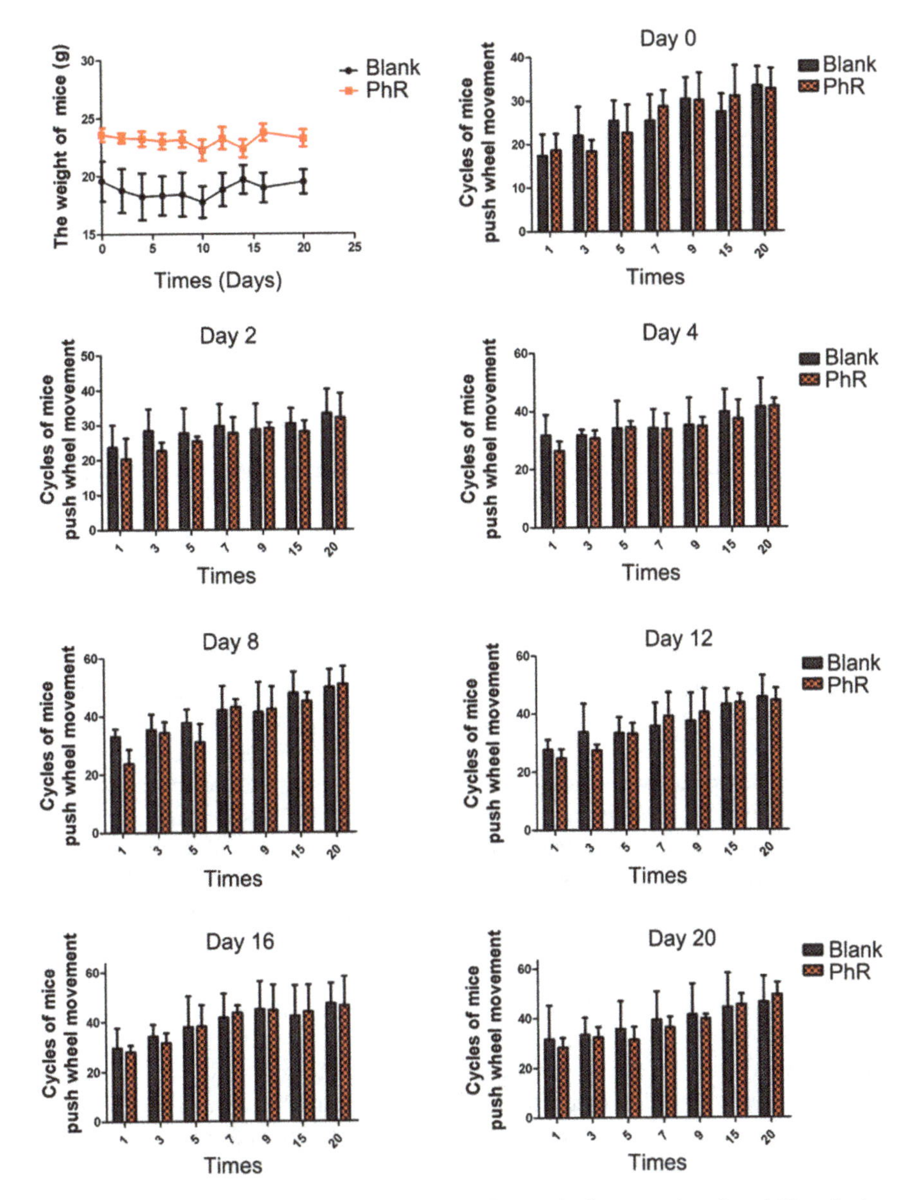


**Fig. 3.23** The cage-wheel exercise assay was used to study the motor learning ability of mice. BALB/c mice were randomly divided into two groups and were subcutaneously injected with PBS or PhR (10 mg/kg). The mice weight kept steady in 20 days after injection. The cycle of mice push wheel in day 20 was no obvious different between control group and PhR group. Over a period of 0–20 days, the weight of mice kept steady and the voluntary running cycles have been increasing steadily and exhibit no significant differences in these two groups, which indicated no obvious effect of motor learning ability of mice treated with PhR. The error bars represent the standard derivations (3 mice per group)

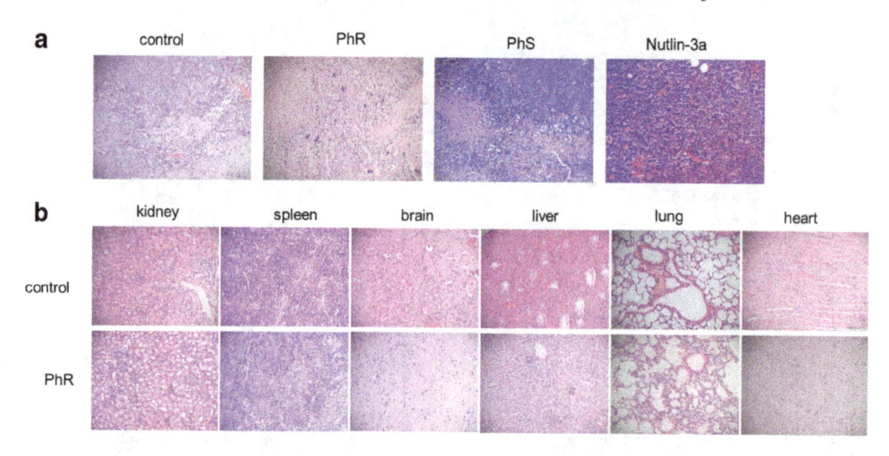


**Fig. 3.24** (**a**) H&E stained tumor sections collected from different groups of mice 3 weeks after treatment. MicroSpot Focusing Objective, 20×. (**b**) H&E stained organs collected from different groups of mice 3 weeks after treatment. Organs collected from two groups of mice (PBS or PhR) 3 weeks post-treatment. No obvious organ hurts were observed. MicroSpot Focusing Objective, 20×

cell (PA-1) xenograft mice model. Notably, very low dosages (10 mg/Kg, every other day injection) were required for PhR to achieve > 70% TGI, in contrast with nutlin-3a's ~30% TGI (10 mg/Kg, one injection every other day). Their efficacy in inhibiting tumor growth in xenograft model at a lower dosage compared with previously described molecules (e.g., small-molecule MDM2-selective inhibitors such as nutlin-3a or the highly potent ATSP-7041) highlights the capacity of CIH peptides as promising preclinical candidates to target intracellular PPIs. Notably, stapled peptide ATSP-7041 only showed minimal in vivo efficacy at 10 mg/Kg dosage [26]. We hypothesize the higher efficacy of PhR may ascribe to the superior nucleus accumulation property and serum stability than stapled peptides (data not shown).

To our knowledge, this is the first report of stabilized peptides showing inhibitory effects on CSCs. Despite its relatively moderate in vitro activity, the superior in vivo efficacy and minimal toxicity of the CIH peptide PhR compared with nutlin-3a shows that it is a promising candidate for further development. From this study, we learned that binding affinity, cellular uptake and function should be evaluated synergetically to provide a suitable candidate, and that in vivo efficacy and toxicity should be examined to obtain a more comprehensive evaluation of a candidate's reactivity.

## 3.4 Methods and Materials

### 3.4.1 Peptide Synthesis

All peptides were synthesized by manual Fmoc-based solid-phase synthesis. The intramolecular thiol-ene reactions were conducted via the method reported in previous literature. The resultant cyclic diastereomers were separated by HPLC. The purified peptides were detected by ESI/LC-MS and the pure fractions were combined and then lyophilized. The detailed synthesis route for PhR was shown below, other peptides synthesis in this article are similar to this route (Schemes 3.1, 3.2 and Table 3.2).

### 3.4.2 Protein Production

Human MDM2 LBD residues 25-117 were cloned into pGEX-4t-1 via EcoRI and XhoI to generate GST-tagged constructs. Expression was carried out in *E. coli* BL21 (DE3) and was induced with 0.1 mM IPTG. Cultures were grown in LB medium at 37°C to an $OD_{600}$ of 0.6 before being transferred to 18°C for 24 h. Cells were harvested by centrifugation and flush frozen. Harvested cells were lysed by sonication in lysis buffer (20 mM Tris-Cl pH 7.9, 500 mM NaCl). Cell debris was removed by centrifugation and the supernatant was purified on a 5 mL GST affinity column (GE healthcare) and eluted with elution buffer (10 mM GSH in 20 mM Tris-Cl pH 7.9, 500 mM NaCl). The protein was further purified with a Superdex 200 column equilibrated in 20 mM Tris-Cl pH 7.9, 500 mM NaCl, 1 mM DTT.

Primer sequences:
MDM2-EcoRI-25: CCGGAATTCGAGACCCTGGTTAGACCAAA
MDM2-XhoI-117: GTAGGCACTCGAGTCAGTCCGATGATTCCT

### 3.4.3 Fluorescence Polarization Assay

FITC-labeled peptides (10–20 nM) were incubated with MDM2 or MDMX protein in binding assay buffer (140 mM NaCl, 50 mM, Tris pH 8.0) at room temperature for 1 h. Fluorescence polarization experiments were performed in 96-well plates (Perkin Elmer Optiplate-96F) on a plate reader (Perkin Elmer, Envision, 2104 multilabel reader). Concentrations of the peptides were determined at 494 nm absorption of FITC. Kd values were determined by nonlinear regression analysis of dose response curves using OriginPro 9.0.

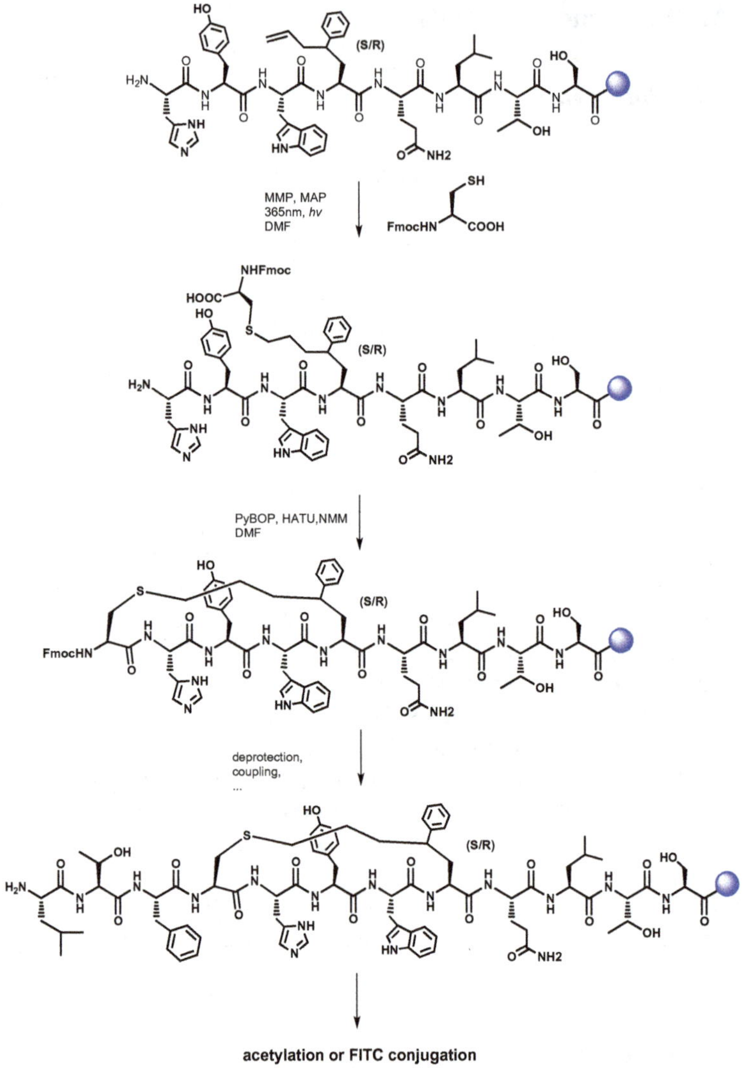

MMP, MAP
365nm, *hv*
DMF

PyBOP, HATU,NMM
DMF

deprotection,
coupling,
...

**acetylation or FITC conjugation**

**Scheme 3.1 Schematic presentation of the synthesis process of peptides**. Here take PhR as an example, the thiol-ene click reaction was performed in a photo-reactor under a wavelength of 365nm violet light. The photoreaction was repeated for three times, each time last for 1hour. The yield of the photoreaction step reached to 90%. The unnatural amino acids were epimer and were synthesized followed the previous reported procedures. For each peptide sequences, we got two diastereomers which we named as S or R. (Abbreviation: MMP: 2-hydroxy-4'-(2-hydroxyethoxy)-2-methylpropiophenone; MAP: 4-methoxyacetophenone; PyBOP: benzotriazol-1-yl-oxytripyrrolidinophosphonium hexafluorophosphate; NMM: N-Methylmorpholine; HATU: N-[(Dimethylamino)-1H-1,2,3-triazolo-[4,5-b]pyridin-1-ylmethylene]-N-methylmethanaminium hexafluorophosphate N-oxide; FITC: fluorescein isothiocyanate; DMF: dimethylformamide)

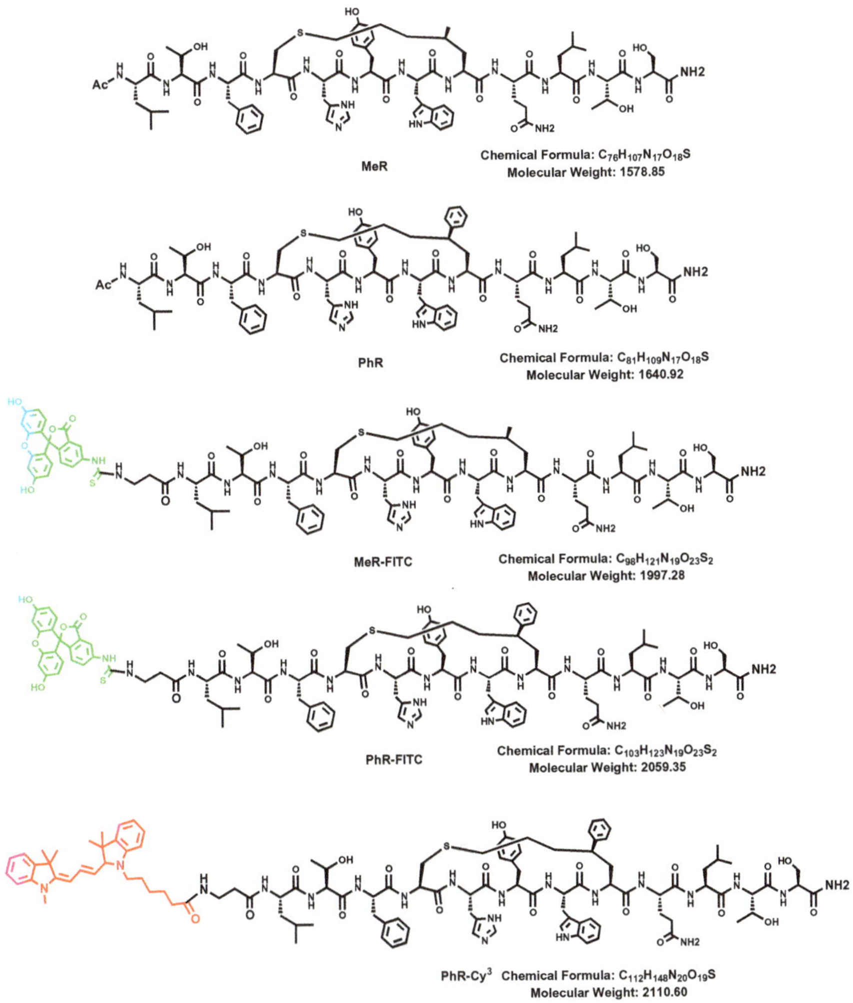

MeR
Chemical Formula: $C_{76}H_{107}N_{17}O_{18}S$
Molecular Weight: 1578.85

PhR
Chemical Formula: $C_{81}H_{109}N_{17}O_{18}S$
Molecular Weight: 1640.92

MeR-FITC
Chemical Formula: $C_{98}H_{121}N_{19}O_{23}S_2$
Molecular Weight: 1997.28

PhR-FITC
Chemical Formula: $C_{103}H_{123}N_{19}O_{23}S_2$
Molecular Weight: 2059.35

PhR-Cy$^3$ Chemical Formula: $C_{112}H_{148}N_{20}O_{19}S$
Molecular Weight: 2110.60

**Scheme 3.2** The chemical structure of PhR/MeR and fluorescein-labeled PhR/MeR

## 3.4.4 Flow Cytometry

1. Transfection efficiency
   For the flow cytometry of transfection efficiency experiments, cells were treated with fluoresceinated peptides (5 μM) for up to 4 h at 37°C. After washing with media, the cells were digested with trypsin (0.25%; Gibco), digestion (2 min, 37°C), washed with PBS, and resuspended in PBS. Cellular fluorescence was analyzed using a BD FACSCalibur flow cytometer (Becton Dickinson) and CellQuest Pro (or CFlow plus). The same experiment was performed with 30 min

**Table 3.2** MS characterization of peptide 1-12

| Entry | | Sequence | | | | | | | | | | | | | | | | Calcu. MS | Obsev. MS |
|---|---|---|---|---|---|---|---|---|---|---|---|---|---|---|---|---|---|---|---|
| 1 | FITC | βA | L | T | F | Q | H | Y | W | A | Q | L | T | S | | | $NH_2$ | 1956.17 | 979.09 |
| **2** | FITC | βA | L | T | F | C | H | Y | W | $S_5$(Me) | Q | L | T | S | | | $NH_2$ | 1999.29 | 1000.03 |
| **3** | FITC | βA | L | T | F | C | H | Y | W | $S_5$(Ph) | Q | L | T | S | | | $NH_2$ | 2061.36 | 1030.2 |
| **4** | FITC | βA | L | T | F | Q | C | Y | W | A | $S_5$(Me) | L | T | S | | | $NH_2$ | 1933.23 | 967.63 |
| **5** | FITC | βA | L | T | F | Q | C | Y | W | A | $S_5$(Ph) | L | T | S | | | $NH_2$ | 1995.3 | 999.02 |
| **6** | FITC | βA | L | C | F | Q | H | $S_5$(Me) | W | A | Q | L | T | S | | | $NH_2$ | 1934.22 | 967.96 |
| **7** | FITC | βA | | | F | C | H | Y | W | $S_5$(Me) | Q | L | T | S | A | A | $NH_2$ | 1927.19 | 964.71 |
| **8** | FITC | βA | L | T | F | C | H | Y | W | $S_5$(Me) | Q | L | T | S | A | A | $NH_2$ | 2141.45 | 1071.75 |
| **9** | FITC | βA | L | T | F | C | H | Y | W | $S_5$(Ph) | Q | L | T | S | A | A | $NH_2$ | 2303.52 | 1152.5 |
| **10** | FITC | βA | L | T | F | Q | C | Y | W | A | $S_5$(Me) | L | T | S | A | A | $NH_2$ | 2075.39 | 1038.99 |
| **11** | FITC | βA | L | T | F | Q | C | Y | W | A | $S_5$(Ph) | L | T | S | A | A | $NH_2$ | 2137.46 | 1070.13 |
| **12** | FITC | βA | L | S | F | C | Q | Y | W | $S_5$(Me) | Cba | L | S | P | | | $NH_2$ | 1970.29 | 986.35 |
| **MeR** | | Ac | L | T | F | C | H | Y | W | $S_5$(R-Me) | Q | L | T | S | | | $NH_2$ | 789.8 | 790.1 |
| **PhR** | | Ac | L | T | F | C | H | Y | W | $S_5$(R-Ph) | Q | L | T | S | | | $NH_2$ | 821.3 | 821.4 |
| **PhR-Cy3** | Cy3 | βA | L | T | F | C | H | Y | W | $S_5$(R-Ph) | Q | L | T | S | | | $NH_2$ | 1056.1 | 1056.2 |

**FITC-labeled peptides were used for FP and cellular uptake, Cy3 labeled peptides were used for in vivo imaging, acetyl peptides were used for all other experiments**

pre-incubation of cells at 4°C followed by 2 hours incubation with fluoresceinated peptides at 4°C to assess the temperature dependence of fluorescent labeling.

2. Cell Cycle
   For the flow cytometry of cell-cycle arrest experiments, cells were treated with peptides and Nutlin-3a for 48 hours. Subsequently, cells were washed twice with phosphate-buffered saline (PBS) and harvested by trypsinization. The cells were fixed with cold 70% ethanol for 4 hours and then underwent centrifugation at 2000 rpm for 5 min to remove the ethanol. Then the fixed cells were dispersed in PBS with 1% Triton-100, 1 mg/mL RNase and 5 mg/mL PI, stained at 37°C for 30 min (double fixation was not needed for this experiment). The samples were analyzed using a FACS Calibur flow cytometer (Becton Dickinson, Mississauga, CA). The percentages of cells in G1, S, and G2/M phases were determined by FlowJo software.
3. Apoptosis assay
   The apoptosis assay was conducted using the FITC Annexin-V/PI Apoptosis Detection Kit I (BD Pharmingen TM) according to the manufacturer's instructions. Approximately $1 \times 10^6$ cells were seeded in a six-well plate. Then, the cells were treated with the peptides and nutlin-3a and incubated for 48 hours. The cells were then collected and washed twice with cold PBS and suspended in binding buffer. The induction of the apoptosis process enables the FITC-labeled Annexin-V to bind with phosphatidylserine (PS) since it appears on the outer surface of the cell membrane at the onset of apoptosis, and is confined to the inner boundary of the cell membrane in healthy cells. The cell nuclei were stained with propidium iodide (PI). The stained cells were analyzed by flow cytometry to distinguish the apoptotic cells. The cells with positive florescent intensity signals for both FITC and PI were representative of the apoptotic cell count. The extent of apoptosis was measured through the detection of caspase-3 activity by exposing the cells to a caspase-3-specific substrate (Oncogene). Fluorescence as a result of substrate cleavage was measured in a Spectramax M5 microplate reader (Molecular Devices).

### *3.4.5 Confocal Microscopy and Co-localization Assay*

PA-1 cells (or MCF-7 cells) were cultured with DMEM with 10% FBS (v/v) in imaging dishes (50000 cells/well) in a 37°C, 5% $CO_2$ incubator for one day until they were about 80% adherent. Peptide was first dissolved in DMSO to make a 1 mM stock and then added to cells to a final concentration of 5 μM. The cells were incubated with peptides for 1 hours at 37°C. After incubation, cells were washed 3 times with PBS and then fixed with 4% formaldehyde (Alfa Aesar, MA) in PBS for 10 minutes. They were then washed 3 times with PBS and stained with 1 μg/ml 4', 6-diamidino-2-phenylindole (DAPI) (Invitrogen, CA) in PBS for 5 minutes. Images of peptide localization in cells were taken via PerkinElmer confocal microscopy. Image processing was done using the Volocity software package (Zeiss Imaging).

### 3.4.6 Cell Viability Assay

Cell viability was measured by the MTT (3-(4,5-dimethylthiazol-2-yl)-2,5-diphenylt-etrazolium bromide, Sigma) assay. Cells were seeded in a 96-well plate at a density of 5 × 103 cells/well and incubated with p53 peptides and nutlin-3a in serum-free media for 4 h, followed by serum replacement and additional incubation for 44 h. MTT (5 mg/mL, 20 μL) in PBS was added and the cells were incubated for 4 h at 37°C with 5% CO2. DMSO (Dimethylsulfoxide, 150 μL, Sigma) was then added to solubilize the precipitate with 5 min of gentle shaking. Absorbance was measured with a microplate reader (Bio-Rad) at a wavelength of 490 nm. Cell viability was obtained by normalizing the absorbance of the sample well against that of the control well and expressed as a percentage, assigning the viability of non-treated cells as 100%.

### 3.4.7 Western Blot Analysis

For western blot analysis, cells were seeded in 6-well plates and treated for 48 h with p53 peptides and nutlin-3a. To isolate the protein, cells were washed with PBS and harvested using lysis buffer (50 mM Tris·Cl PH = 6.8, 2% SDS, 6% Glycerol, 1% β-mercapitalethanol, 0.004% bromophenol blue). Total cellular protein concentrations were determined by a spectrophotometer (Nano-Drop ND-2000). 20 μg of denatured cellular extracts were resolved using 10% SDS-PAGE gels. Protein bands in the gel were then transferred to nitrocellulose blotting membranes and incubated with the appropriate primary antibody. The antibody dilutions were as follows: 1:500 for MDM2, MDMX, p53 and 1:1000 for actin, H3. Membranes were incubated overnight at 4°C and washed the next day with buffer (1 × PBS, 0.05% Tween 20). Goat anti-rabbit or anti-mouse secondary antibodies were used for secondary incubation for 1 hours at room temperature. Proteins were then visualized with chemiluminescent substrates.

### 3.4.8 RNA Extraction and RT-PCR and Microarray Analysis

For western blot analysis, cells were seeded in 6-well plates and treated for 48 h with P53 peptides and nutlin-3a as described for the western blot assay. Then total RNA was extracted from cells using TRIzol reagent (Invitrogen) and the amount of RNA was quantified by a spectrophotometer (Nano-Drop ND-2000). Total RNA (2 μg) was reverse transcribed to cDNA using the reverse transcriptase kit from Promega according to the manufacturer's instructions. The mRNA levels of the target genes were quantified by real-time PCR using SYBR green (Promega) in an ABI Prism 7500 real-time PCR system (Applied Biosystems). The primers used are listed in

**Table 3.3** Primer sequence of genes for RT-PCR analysis

| Primer name | Sequence (5'-3') |
|---|---|
| Forward-P53 | GGAGCACTAAGCGAGCACTG |
| Reverse-P53 | TATGGCGGGAGGTAGACTGA |
| Forward-MDM2 | GGGCTTTGATGTTCCTGATT |
| Reverse-MDM2 | CTTTGTCTTGGGTTTCTTCC |
| Forward-MDMX | CATTTCGGCTCCTGTCGTTA |
| Reverse-MDMX | GTTCCCGTCTCGTGGTCTTT |
| Forward-MIC1 | AGTTGCGGAAACGCTACGAG |
| Reverse-MIC1 | GGAACAGAGCCCGGTGAAGG |
| Forward-Sox2 | GTGAGCGCCCTGCAGTACAA |
| Reverse-Sox2 | GCGAGTAGGACATGCTGTAGGTG |
| Forward-FoxA2 | CCCCAACAAGATGCTGACGC |
| Reverse-FoxA2 | GCGAGTGGCGGATGGAGTT |
| Forward-beta Actin | TCCAGCCTTCCTTCTTGGGTATG |
| Reverse-beta Actin | GAAGGTGGACAGTGAGGCCAGGAT |

Table 3.1. For microarray analysis, the PA-1 cells were treated with PhR (40 μM) for 48 hours, the total RNA were extracted. Labeled RNA was hybridized to microarrays (Human HT-12 BeadChip; Illumina). Raw signal intensities of each probe were obtained using data analysis software (Beadstudio; Illumina) and imported to the Lumi package of Bioconductor for data transformation and normalization. Differentially expressed genes were identified using ANOVA model with empirical Bayesian variance estimation. The problem of multiple comparisons was corrected using the false discovery rate (FDR). Genes were identified as differentially expressed based on statistical significance (raw P value < 0.05), FDR < 0.1%, and > 2-fold change (up or down) in expression level. The microarray and data analysis was completed in Beijing Genomics Institute (BGI Inc.) (Table 3.3).

### *3.4.9 Ubiquitination Analysis*

For the cytoplasmic translocation assay, HCT116 cells were plated on Costar-six-well-containing glass coverslips until they reached 80–90% confluences. The following day, the cells were transfected with GFP-p53 (1 μg) using HD transfection reagent (Roche, USA) according to the manufacturer's protocol (for low levels of Mdm2 expression, 0.5 μg of pCMV-Mdm2 were used). At 4 h after transfection, cells were treated with peptide. Twenty-four hours after transfection, cells on the coverslips were washed three times with phosphate-buffered saline (PBS) and then fixed in 4% paraformaldehyde/PBS for 10 min at room temperature. After 3 washes with ice-cold PBS, cells were perforated in ice-cold PBS containing 0.2% Triton X-100 for 10 min. Cells were blocked in PBS containing 1% bovine serum albumin and 1 μg of DAPI (Sigma)/ml at room temperature for 30 min in a dark environment. Cells were washed three times with PBS, and the stained cells were mounted with mounting medium and the coverslips were sealed with nail polish. Fluorescence was recorded using a confocal microscope.

### 3.4.10 *Preparation of Paraffin Section Histological Analysis (IHC)*

For histological experiments, organ tissues were collected on the final day treatment and fixed in 4% buffered formalin-saline at room temperature for 24 hours. Following this, tissues were embedded in paraffin blocks and 4 mm thick paraffin sections were mounted on a glass slide for hematoxylin and eosin (H&E) staining. The H&E staining slices were examined under a light microscopy (Olympus BX51).

### 3.4.11 *Antitumor Efficacy in Human Xenograft Model Using PhR Peptide*

Athymic nude mice (BALB/c ASlac-nu) were obtained from Vital River Laboratory Animal Technology Co. Ltd. of Beijing, People's Republic of China and allowed an acclimation period of 1 week. Mice were maintained in an isolated biosafety facility for specific-pathogen-free (SPF) animals with bedding, food, and water. All operations were carried out in accordance with the National Standard of Animal Care and Use Procedures at the Laboratory Animal Center of Shenzhen University, Guangdong Province, People's Republic of China (the permit number is SZU-HC-2014-02). For tumor suppression assay, athymic nude mice (female; 6 weeks old) were inoculated with $1 \times 10^7$ PA-1 cells (PA-1 cells were trypsinized, harvested, and resuspended in DMEM, with 100 mL volume of each) propagated in vitro subcutaneously in the lower flank of mice. After 10–15 days, mice with tumors exceeding 100–150 mm$^3$ in volume were randomly divided into 3 groups of 5–6 mice per treatment group. Mice bearing PA-1 tumors were injected with PhR (10 mg/Kg) or nutlin-3a (10 mg/Kg) using PBS as negative control. Mice were injected once every 2 days starting on day 0. Tumor volumes were measured by calipers (accuracy of 0.02 mm) every other day and calculated using the following formula: $V = L \times W^2/2$ (W, the shortest dimension; L, the longest dimension). Each tumor was independently measured and calculated by changes in volume (folds) relative to day 0. Statistical significances between groups were tested by one-way analysis of variance.

### 3.4.12 *Mice Voluntary Cage-Wheel Exercise*

BALB/c mice (female; 6 weeks old) were obtained from Vital River Laboratory Animal Technology Co. Ltd. of Beijing, People's Republic of China and allowed an acclimation period of 1 week at 22 ± 2 °C with a 12-hour light: dark cycle (lights on at 8am, lights off at 8 pm). Subsequently, BALB/c mice were randomly divided into 2 groups (3–4 mice per group) and were subcutaneously injected with PBS or peptide (10 mg/kg). Voluntary running was performed by these two groups at the start of exercise following reported protocol. A voluntary running system consisting of six separated chambers (Chengdu TME Technology Co., Ltd, China) was used in the animal performance study. During the training session, mice were placed on the motorized rod (30 mm in diameter) in the chamber. The rotation speed gradually increased from 0 to 100 rpm over the course of 100 s. The rotation speed was recorded when the animal fell off from the rod. Each rotarod training session consisted of 7 trials and lasted around 7 minutes. Performance was measured as the average rotation speed animals achieved during the training session. The two different groups were all trained at the same five continuous time points (day 0, day 2, day 4, day 8, day 12, day 16, and day 20). No significant differences were found between the two groups.

### 3.4.13 *In Vivo Imaging*

When the tumor reached an appropriate volume of 200–300 $mm^3$, the mice were injected with 100 uL of Cy3-labeled PhR peptide by intratumoral injection. After injection, mice were anesthetized with isoflurane. The induction concentration was 5% isoflurane/1L $O_2$, and the maintenance concentration was 2–3% isoflurane/1L $O_2$. Once the mice were properly anesthetized, they were imaged at indicated time points to monitor the metabolization of $PhR^{Cy3}$ peptide in tumors using the IVIS LuminalI small animal in vivo optical imaging system (Caliper). In this study, a scanning wavelength ranging between 500 and 950 nm was used for in vivo imaging.

## References

1. Frank NY et al (2010) The therapeutic promise of the cancer stem cell concept. J Clin Investig 120(1):41–50
2. Ni C, Huang J (2013) Dynamic regulation of cancer stem cells and clinical challenges. Clin Transl Oncol 15(4):253–258
3. Marhold M et al (2015) HIF1α regulates mTOR signaling and viability of prostate cancer stem cells. Mol Cancer Res 13(3):556–564
4. Li Y, Laterra J (2012) Cancer stem cells: distinct entities or dynamically regulated phenotypes? Can Res 72(3):576–580
5. Chen K et al (2013) Understanding and targeting cancer stem cells: therapeutic implications and challenges. Acta Pharmacol Sin 34(6):732–740

6. Liu J et al (2009) Biorecognition and subcellular trafficking of HPMA copolymer—anti-PSMA antibody conjugates by prostate cancer cells. Mol Pharm 6(3):959–970
7. Jin L et al (2006) Targeting of CD44 eradicates human acute myeloid leukemic stem cells. Nat Med 12(10):1167–1174
8. Liu F-S (2009) Mechanisms of chemotherapeutic drug resistance in cancer therapy—a quick review. Taiwan J Obstet Gynecol 48(3):239–244
9. Moitra K et al (2011) Multidrug efflux pumps and cancer stem cells: insights into multidrug resistance and therapeutic development. Clin Pharmacol Ther 89(4):491–502
10. Moellering RE et al (2009) Direct inhibition of the NOTCH transcription factor complex. Nature 462(7270):182–188
11. Bird GH et al (2010) Hydrocarbon double-stapling remedies the proteolytic instability of a lengthy peptide therapeutic. Proc Natl Acad Sci 107(32):14093–14098
12. Milroy L-G et al (2014) Modulators of protein-protein interactions. Chem Rev 114(9):4695–4748
13. Hu K et al (2016) An in-tether chiral center modulates the helicity, cell permeability, and target binding affinity of a peptide. Angew Chem Int Ed 55(28):8013–8017
14. Walensky LD et al (2004) Activation of apoptosis in vivo by a hydrocarbon-stapled BH3 Helix. Science 305(5689):1466–1470
15. Bernal F et al (2007) Reactivation of the p53 tumor suppressor pathway by a stapled p53 peptide. J Am Chem Soc 129(9):2456–2457
16. Grossmann TN et al (2012) Inhibition of oncogenic Wnt signaling through direct targeting of β-catenin. Proc Natl Acad Sci 109(44):17942–17947
17. Vogelstein B et al (2000) Surfing the p53 network. Nature 408(6810):307–310
18. Haupt S et al (2003) Apoptosis—the p53 network. J Cell Sci 116(20):4077–4085
19. Cheok CF et al (2011) Translating p53 into the clinic. Nat Rev Clin Oncol 8(1):25–37
20. Hu B et al (2006) MDMX overexpression prevents p53 activation by the MDM2 inhibitor nutlin. J Biol Chem 281(44):33030–33035
21. Mogi A, Kuwano H (2011) TP53 mutations in nonsmall cell lung Cancer. J Biomed Biotechnol 2011
22. Obrador-Hevia A et al (2015) RG7112, a small-molecule inhibitor of MDM2, enhances trabectedin response in soft tissue sarcomas. Cancer Invest 33(9):440–450
23. Tovar C et al (2013) MDM2 small-molecule antagonist RG7112 activates p53 signaling and regresses human tumors in preclinical cancer models. Can Res 73(8):2587–2597
24. Graves B et al (2012) Activation of the p53 pathway by small-molecule-induced MDM2 and MDMX dimerization. Proc Natl Acad Sci 109(29):11788–11793
25. Brown CJ et al (2013) Stapled peptides with improved potency and specificity that activate p53. ACS Chem Biol 8(3):506–512
26. Chang YS et al (2013) Stapled α − helical peptide drug development: a potent dual inhibitor of MDM2 and MDMX for p53-dependent cancer therapy. Proc Natl Acad Sci U S A 110(36):E3445–E3454
27. Chung W-M et al (2013) MicroRNA-21 promotes the ovarian teratocarcinoma PA1 cell line by sustaining cancer stem/progenitor populations in vitro. Stem Cell Res & Ther 4(4):1–10
28. Chung W-M et al (2014) Ligand-independent androgen receptors promote ovarian teratocarcinoma cell growth by stimulating self-renewal of cancer stem/progenitor cells. Stem Cell Res 13(1):24–35
29. Sekar D et al. Deciphering the role of microRNA 21 in cancer stem cells (CSCs). Genes & Diseases
30. Yaginuma Y, Westphal H (1992) Abnormal structure and expression of the p53 gene in human Ovarian carcinoma cell Lines. Can Res 52(15):4196–4199
31. Reich NC et al (1983) Two distinct mechanisms regulate the levels of a cellular tumor antigen, p53. Mol Cell Biol 3(12):2143–2150
32. Wang L et al (2001) Analyses of p53 target genes in the human genome by bioinformatic and microarray approaches. J Biol Chem 276(47):43604–43610

33. Hu B et al (2007) Efficient p53 activation and apoptosis by simultaneous disruption of binding to MDM2 and MDMX. Can Res 67(18):8810–8817
34. Baek S et al (2012) Structure of the stapled p53 peptide bound to Mdm2. J Am Chem Soc 134(1):103–106
35. Chen L et al (1999) Ubiquitous induction of p53 in tumor cells by antisense inhibition of MDM2 expression. Mol Med 5(1):21–34
36. Ventura A et al (2007) Restoration of p53 function leads to tumour regression in vivo. Nature 445(7128):661–665
37. Wang W et al (2003) Stabilization of p53 by CP-31398 inhibits ubiquitination without altering phosphorylation at serine 15 or 20 or MDM2 binding. Mol Cell Biol 23(6):2171–2181
38. Brooks H et al (2005) Tat peptide-mediated cellular delivery: back to basics. Adv Drug Deliv Rev 57(4):559–577
39. Zhang X et al (2013) Pluripotent stem cell protein Sox2 confers sensitivity to LSD1 inhibition in cancer cells. Cell Reports 5(2):445–457
40. Wahl AF et al (1996) Loss of normal p53 function confers sensitization to taxol by increasing G2/M arrest and apoptosis. Nat Med 2(1):72–79
41. Ryan KM et al (2001) Regulation and function of the p53 tumor suppressor protein. Curr Opin Cell Biol 13(3):332–337
42. Hsieh J-K et al (1999) RB regulates the stability and the apoptotic function of p53 via MDM2. Mol Cell 3(2):181–193
43. Huang B et al (2009) Pharmacologic p53 activation blocks cell cycle progression but fails to induce senescence in epithelial cancer cells. Mol Cancer Res 7(9):1497–1509
44. Pishas KI et al (2011) Nutlin-3a is a potential therapeutic for ewing sarcoma. Clin Cancer Res 17(3):494–504
45. Brooks CL, Gu W (2006) p53 ubiquitination: Mdm2 and beyond. Mol Cell 21(3):307–315
46. Paek Andrew L et al (2016) Cell-to-cell variation in p53 dynamics leads to fractional killing. Cell 165(3):631–642
47. Kelly PN et al (2007) Tumor growth need not be driven by rare cancer stem cells. Science 317(5836):337–337
48. Clevers H (2011) The cancer stem cell: premises, promises and challenges. Nat Med, 313–319
49. Visvader JE, Lindeman GJ (2008) Cancer stem cells in solid tumours: accumulating evidence and unresolved questions. Nat Rev Cancer 8(10):755–768
50. Takebe N et al (2011) Targeting cancer stem cells by inhibiting Wnt, notch, and hedgehog pathways. Nat Rev Clin Oncol 8(2):97–106
51. Vermeulen L et al (2010) Wnt activity defines colon cancer stem cells and is regulated by the microenvironment. Nat Cell Biol 12(5):468–476
52. Dean M (2009) ABC transporters, drug resistance, and cancer stem cells. J Mammary Gland Biol Neoplasia 14(1):3–9
53. Liu C et al (2011) The microRNA miR-34a inhibits prostate cancer stem cells and metastasis by directly repressing CD44. Nat Med 17(2):211–215
54. Holohan C et al (2013) Cancer drug resistance: an evolving paradigm. Nat Rev Cancer 13(10):714–726
55. Watt FM et al (2000) Out of eden: stem cells and their niches. Science 287(5457):1427–1430
56. Li L, Neaves WB (2006) Normal stem cells and cancer stem cells: the niche matters. Can Res 66(9):4553–4557

# Chapter 4
# Summary and Conclusion

Peptide medicine has experienced nearly a hundred years of history from the earliest application of bovine insulin to nowadays. Over the past century, with the continuous advancement of peptide synthesis methodology and purification methods, new peptide molecules have entered the medical market one after another and several of them have become blockbuster drugs for disease treatment. However, it should not be ignored that most of these peptide molecules are obtained from natural products or modified products of natural molecules. There are very few examples of artificially designed peptide molecules becoming drugs. With the continuous development of structural biology and systems biology, new drug targets are constantly being revealed. To design effective ligands for these targets, small molecule shows limitations. The emergence of new molecular modalities of drugs is continuing. Peptides are considered to be the most effective molecular form for targeting non-druggable PPIs. Therefore, designing effective peptide drugs, enhancing the specificity of peptide molecules, improving the penetrating ability of peptide molecules, and reducing the non-specific toxicity of peptide molecules are important goals for the development of peptide drugs in the future.

This thesis focuses on the construction of stabilized helical peptides and creatively proposed a novel chirality-induced helix system, from methodological development to structure-activity relationship research, and further develops protein-protein interaction inhibitors based on this method to verifies the utility of the method.

To organize this thesis, I always focus on the most pressing issues in peptide medicinal chemistry, trying to theoretically address some of the most fundamental questions in the field of stabilized helical peptides, and put forward some basic principles to instruct peptide drug design.

Chapter 1 introduces the importance of protein-protein interactions and the advantages of peptides in targeting protein-protein interactions. Moreover, a brief overview of the current methodology for constructing helical peptides and the associated limitations of them is performed.



K. Hu, *Development of In-Tether Carbon Chiral Center-Induced Helical Peptide*, Springer Theses,
https://doi.org/10.1007/978-981-33-6613-8_4

Chapter 2 introduces the development of the CIH strategy. The addition of a precisely positioned chiral center in the tether of a constrained peptide is shown to yield two separable peptide isomers with significantly different helical content. These results are validated by circular dichroism, NMR spectroscopy, X-ray crystallography, and molecular dynamics simulation. Single crystal X-ray analysis reveals that the absolute configuration of the chiral center in the helical form is *R*. The peptide helicity could be further enhanced by increasing the bulkiness of the (*R*)-chiral center substitution group. Molecular dynamics simulations using a recently developed RSFF1 force field were employed to rationalize the above observations and explained the origin of the in-tether chirality-induced helicity of the peptides. The CIH strategy provides an ideal platform to investigate the solely conformational influence on peptides' biophysical properties. Further studies reveal that the helical (*R*)-peptides display a better cell permeability than the corresponding (*S*)-peptides. Notably, the (*R*) isomers of long bioactive peptides also showed significantly better α-helicity contents, binding affinity, and cell permeability than their (*S*) isomers. These results unambiguously demonstrated the close correlation between peptides' helicity and their biophysical properties.

Chapter 3 introduces the application of CIH strategy in designing inhibitors for P53-MDM2/X interactions. Inhibition of the interaction between P53 and MDM2/MDMX has attracted significant attention in anticancer therapy development. We designed a series of in-tether chiral center-induced helical stabilized peptides, among which MeR/PhR effectively reactivated P53. The activation of P53 inhibits cell proliferation and induces apoptosis in both the MCF-7 normal tumor cell line and the PA-1 pluripotent cancer cell line with only minimal cellular toxicity toward normal cells or cancer cell lines with P53 mutations. The in vivo bioactivity study of the peptide in the ovarian teratocarcinoma (PA-1) xenograft model showed a tumor growth rate inhibition of 70% with a dosage of 10 mg/kg (one injection every other day). Notably, due to the pluripotent nature of the PA-1 cell line, the small molecular inhibitor nutlin-3a showed only very limited effects in vivo. Its low toxicity and long duration time in vivo render our peptide as a potential drug candidate for teratocarcinoma therapy. Significantly, this is the first application of a stabilized peptide modulator targeting stem-like cancer cells both in vitro and in vivo and provides references to cancer stem cell therapy.

In summary, this thesis focuses on the chirality-induced helix system, starting from basic research of how an in-tether chiral center influences the secondary structure of a peptide. Next, I conducted systematic research on the structure-activity relationship and druggability of several kinds of bioactive peptides synthesized by using this method. I hope that through this series of studies, the peptide science community will have a better understanding of the structure and function of helical peptides and help the scientific community develop the new generation of peptide drugs.

CPSIA information can be obtained
at www.ICGtesting.com
Printed in the USA
LVHW080431100221
678688LV00001BB/4